Chernobyl

01:23:40

by Andrew Leatherbarrow

Published by Andrew Leatherbarrow

Published by Andrew Leatherbarrow

"Imagine personnel of a plane which is flying very high. Whilst flying they begin testing the plane, opening the doors of the plane, shutting off various systems… The facts show that even such a situation should have been foreseen by the designers."

Valeri Legasov,
USSR Delegation Leader, 25-29th August 1986, Vienna.

Preface

When I first began reading books about Chernobyl, I found them very difficult to follow. My first, *Chernobyl Notebook* by Grigoriĭ Medvedev, a Soviet nuclear power station inspector, while an excellent book, assumed a great deal of prior knowledge of nuclear systems and the translation was raw. Over time - and books - I grew more at-ease with the technology and terminology, but still felt a lingering concern that these books were just too difficult for the average person to jump into. The Chernobyl disaster is one of the most incredible and globally significant events of the last 100 years, yet so few people understand what happened.

This confusion partly arose because all information released during the five years following the accident was distorted to fit a chosen narrative – that the power station staff were to blame. From there, bits and pieces of drip-fed information evolved into myths and legends, despite later clarification of many of the earlier inaccuracies. Every new book, documentary, newspaper and website told a slightly different version of the story, and contradictions are still common today. Not only that, but I couldn't find a single title that focused on the parts of the story I was most interested in. Most dedicate only a small section to the actual accident itself, choosing instead to focus on the aftermath. Those that do detail the actual event, such as *Chernobyl Notebook*, tend to ignore the aftermath entirely. Others become embroiled in politics, environmentalism or endless numbers. So, after looking for the sort of book I wanted to read and discovering that no such title existed, I decided to write one myself.

I do not want to sensationalise the accident. What happened *is* sensational in many ways, but the story is frequently exaggerated for the sake of adding drama. It's dishonest and unnecessary; the true events were dramatic enough. Nor have I set out to condemn or absolve anyone. I can't stand it when authors of non-fiction force their opinions on the reader, I'm simply presenting the facts as I see them.

While I have made painstaking efforts to ensure that the details presented here are correct, certain aspects - mostly relating to the reactor - have been deliberately simplified to an extent, for the sake of keeping the story easy to follow. I have also kept the number of characters and their specific stories to a minimum for the sake of brevity, concentrating on those whom I feel were most central to what took place. I also felt it was important to make this as relatable a story as possible, and hence have used a lot of quotes from people who were there. Over time, I have been forced to conclude that it will never be possible to create a 100% accurate account of what happened because of conflicting witness information, but I have gone far out of my way to ensure that this book is as truthful as it can be. Where I'm unsure of something, I have noted it in the footnotes.

I wanted to include an account of my 2011 journey to Chernobyl that prompted my desire to study the disaster further.

It was a most profound experience for me and changed my life completely. This second narrative, while clearly less enthralling than the historical account, breaks up the book and hopefully adds something to the overall package. I don't remember many of the finer points or conversations of this trip and didn't want to make up information to pad it out, so the lack of detail in some areas is a deliberate choice. All of the Pripyat and Chernobyl photographs included with this book are taken by me during this trip. The entire collection of about 1000 photos can be found here: https://goo.gl/uchbWp

It's taken me four and a half years and thousands of hours of writing and research, entirely in my spare time, to reach this point. For the first two years or so I had no intention of releasing this as an actual book. I was writing it for myself as a hobby, to perhaps print one copy and put it on a shelf. As such, I made the amateur mistake of not keeping a record of my sources up until that point, so I've had to go back and re-find a great deal of information when assembling the references. Because of this, some of the references included here aren't necessarily where I first found the information. While working on the text, I made it all available online for free and updated it over time as it grew longer and longer. It was only once I began receiving emails urging me to release it as a proper book that I started to entertain the notion. I set up a Kickstarter in an effort to raise funding for an editor in early 2015, but that fell flat on its face, at which point I abandoned the project entirely.

In April of the same year, I uploaded an album of 150 historic photographs of Chernobyl to Reddit in honour of the disaster's 29th anniversary, complete with captions from my book. The reception was astonishing. People asked me to make the book available in its then-current state, and so for two days I did. Within an hour, I had the book uploaded to a print-on-demand website and in those two days I somehow sold over 700 copies. Me, a nobody with zero credentials. This proved to me that people were interested in the disaster.

Five weeks later my first baby boy, Noah, was born, and I left Chernobyl behind for a while. By September I decided it was stupid to abandon the book when it was so close to completion. With no money to pay a professional, I found some editing software and edited it myself. The months off gave me time to see areas that needed extra detail and I received a lot of invaluable feedback from the people of Reddit who bought my unedited book. I made changes accordingly and have no doubt that the final product is all the better for it.

The book was finished in March 2016, after six sleep-deprived months (thanks Noah) of editing in my spare time. Then, amazingly, a young editor from Reddit named Elizabeth Petrey found my work and offered to help for free. She has done a wonderful job racing through it during these final few weeks. Reddit has proven to be an invaluable help along the way. From nuclear engineers correcting my physics to college historians correcting my history to Russians correcting my translations, I really am indebted to the wonderful people from that website and cannot thank them enough.

I am no writer, certainly not in the traditional sense. I have had no training of any kind and had never written anything at all before attempting this project. My first drafts were awful and I have rewritten the entire thing more times than I want to remember, but over time I got (slightly) better at it. I'm the first to admit that this is far from the best thing I've ever read, but it's as good as I can manage at the moment and I hope you enjoy it.

Finally, I would like to state for the record that I am in favour of nuclear power in developed nations, when strict health, safety and environmental considerations are adhered to.

Contents

CHAPTER ONE

A Brief History of Nuclear Power

Radiation is perhaps the most misunderstood phenomenon known to humanity. Even today, now that its effects are well known, the word 'radiation' still elicits a fearful overreaction in most people. During the euphoric decades of study following its discovery at the turn of the century, people held a more carefree attitude in their ignorance. Radiation's most well-known pioneering researcher, Marie Curie, died in 1934 from aplastic anaemia brought on by her decades of unprotected exposure to the faint, glowing substances in her pockets and desk drawers. Together with her husband Pierre, she built upon German physicist Wilhelm Röntgen's momentous 1895 discovery of X-rays, by working tirelessly out of, *"an abandoned shed which had been in service as a dissecting room of the School of Medicine,"*[1] on the University of Paris' grounds. Curie herself noted that, *"one of our joys was to go into our workroom at night... the glowing tubes looked like*

---

[1] Curie, Marie. *Pierre Curie, with the Autobiographical Notes of Marie Curie*. New York: Dover Publications, 1963. Page 91.

faint fairy lights."[2] While researching the chemical element Uranium, the pair discovered and named new elements Thorium, Polonium and Radium, and spent significant time studying the effects of unusual waves radiating from all four. Marie dubbed these waves 'radiation' and received the Nobel Prize for her work. Until this point in time, the atom was believed to be the absolute smallest thing in existence. It was accepted that atoms were whole, unbreakable, and by themselves formed the building blocks of the universe. Curie's revelation that radiation is created when atoms split apart was groundbreaking.

Her discovery that the fluorescent Radium destroyed diseased human cells faster than it destroyed healthy cells spawned a whole new industry in the early 20th century, peddling the (mostly imagined) properties of this magical new element to an unsuspecting and misguided public. This craze was encouraged by authority figures, including a Dr. C. Davis, who wrote in the American Journal of Clinical Medicine that, *"Radioactivity prevents insanity, rouses noble emotions, retards old age, and creates a splendid, youthful, joyous life."*[3] Watch and clock faces, fingernails, military instrument panels, gun sights and even children's toys glowed with radium, hand-painted in factories by young women working for the United States Radium Corporation. The unsuspecting artisans would lick their brushes - ingesting radium particles each time - to keep the tips pointed during the precision work, but years later their teeth and skulls began to disintegrate. *Radithor,* a 'modern weapon of curative science' and one of several medicinal radium products of the time, boasted that it could cure people of rheumatism, arthritis

---

[2] Ibid. Page 92.
[3] Wells, Jonathan C. K., S. S. Strickland, and Kevin N. Laland. *Social Information Transmission and Human Biology.* Boca Raton, FL: CRC/Taylor & Francis, 2006. Page 150.

and neuritis.[4] Radium cosmetics and toothpastes promising to rejuvenate the skin and teeth were popular for a few years, as were various other proud-to-be-radioactive products, such as radium condoms; chocolate; cigarettes; bread; suppositories; wool; soap; eye drops; *The Scrotal Radiendocrinator* (from the same genius who brought us *Radithor*) to enhance a man's virility; and even radium sand for children's sandpits, advertised by its creator as, *"most hygienic and… more beneficial than the mud of world-renowned curative baths."*[5] The true hazardous properties of radium, which is around 2.7 million times more radioactive than uranium, were not realised or acknowledged by the public until the 1930s and 40s.[6][7]

Fevered work to uncover the atom's secrets continued throughout the early years of the 20th century, as scientists across Europe made many important breakthroughs. By 1932, American physicist James Chadwick made his Nobel winning discovery of the neutron - the final missing piece of the puzzle. With Chadwick's discovery, the atom's structure had been unlocked: an atom consists of a nucleus - a central region of protons and neutrons - circled by electrons. The atomic age had truly begun.

Several years later in 1939, physicists Lise Meitner, Otto Frisch and Niels Bohr determined that when an atom nucleus splits and creates new nuclei (a process called nuclear fission), it releases vast amounts of energy, and that a fission chain reaction was possible. The news brought with it the theory that such a

---

[4] *"Radithor (ca. 1925-1928)."* Oak Ridge Associated Universities. February 17, 2009. Accessed February 24, 2016. https://www.orau.org/ptp/collection/quackcures/radith.htm.
[5] Clark, Claudia. *Radium Girls, Women and Industrial Health Reform: 1910-1935.* Chapel Hill, NC: University of North Carolina Press, 1997. Chapter 1.
[6] Malley, Marjorie Caroline. *Radioactivity: A History of a Mysterious Science.* New York: Oxford University Press, 2011. Page 115.
[7] Orci, Taylor. *"How We Realized Putting Radium in Everything Was Not the Answer."* The Atlantic. March 7, 2013. http://www.theatlantic.com/health/archive/2013/03/how-we-realized-putting-radium-in-everything-was-not-the-answer/273780/.

chain reaction could potentially be harnessed to create a limitless supply of clean energy for ships, planes, factories and homes, or unleashed from a weapon of unfathomable destructive force. Just two days before World War II began, Bohr and John Wheeler published a paper proposing that fission would work better in an environment where a 'moderator' was introduced to slow the speed of neutrons moving within an atom, thus giving them a greater chance to collide and split away from one another.[8]

As the dangers of radioactive products became more well-known and their civilian popularity collapsed, the desperation and urgency of World War II brought about other remarkable advances in the field. Britain was initially the country most devoted to unlocking the secrets of a fission weapon. Germany had a nuclear program, but it focused on power reactor development. After the Japanese attack on Pearl Harbor on December 7th, 1941, America - which had previously concentrated on nuclear naval propulsion - began its own serious fission research, applying vast resources to the development of an atomic bomb. Within a year, the world's first nuclear reactor, Chicago Pile-1, was built at the University of Chicago as part of America's Manhattan Project, supervised by Nobel Laureate for Physics Enrico Fermi. The reactor, famously described by Fermi as, *"a crude pile of black bricks and wooden timbers,"* first went critical (achieved a self-sustaining chain reaction) on December 2nd, 1942.[9] Using graphite as its moderator, the reactor had neither a radiation shield nor a cooling system of any kind.[10] It was a massive and reckless risk

[8] Rhodes, Richard. *The Making of the Atomic Bomb*. New York: Simon & Schuster, 1986.
[9] Wood, J. *Nuclear Power*. London: Institution of Engineering and Technology, 2007. Page 10.
[10] Rhodes, Richard. *The Making of the Atomic Bomb*. New York: Simon & Schuster, 1986. Page 436

by Fermi, who had to convince his colleagues that his calculations were accurate enough to rule out an explosion.

Joseph Stalin learned that the United States, Britain and Germany were all pursuing fission after a physicist named Georgi Flerov, returning from the front lines, noticed all research on nuclear physics had disappeared from the newly published international science journals. The young man (who now has an artificial chemical element named after him: Flerovium) realised the articles had become classified and wrote a letter to Stalin, in which he stressed the significance of their absence; *"build the uranium bomb without delay."*[11] The dictator took notice and devoted more resources towards the potential of fission energy. He instructed prominent Russian scientist Igor Kurchatov to focus on coordinating espionage information on the Manhattan Project, and to begin surreptitious research into what would be necessary for the Soviets to build a bomb. To do this in absolute secrecy, Kurchatov established a new laboratory, hidden away in Moscow's wooded outskirts.

The Allied forces declared victory over Germany on May 8th, 1945, and America turned its attention to Japan. Meanwhile, Kurchatov had made rapid progress but was still behind the Americans, who, under Robert Oppenheimer, successfully tested the first atomic device at 05:29:21 on July 16th, 1945, near Alamogordo, New Mexico.[12] As this was the first time a weapon of such devastating potential had been tested and the consequences were unproven, Fermi offered to take wagers from the physicists and military personnel present as to whether the bomb would ignite the atmosphere, and, if it did, would it

---

[11] Cochran, Thomas B., Robert S. Norris, and Oleg Bukharin. *Making the Russian Bomb: From Stalin to Yeltsin.* Boulder, CO: Westview Press, 1995. Page 19.
[12] Gutenberg, B. *"Interpretation of Records Obtained from the New Mexico Atomic Test, July 16, 1945."* Report. Bulletin of the Seismological Society of America. ISSN 0037-1106. 1945.

only destroy the state or the entire planet.[13] Codenamed 'Trinity', the blast dug a crater 1,200 feet in diameter, and produced temperatures of 'tens of millions of degrees Fahrenheit'. Frightened by what he had witnessed, physicist George Kistiakowsky said, *"I am sure that at the end of the world, in the last millisecond of the Earth's existence, the last man will see what we have just seen."*[14] A mere three weeks later, on August 6th, a modified Boeing B-29 Superfortress dropped the first atomic bomb on the city of Hiroshima, Japan, and its population of 350,000 people. It converted 0.6 grams of uranium into a force of energy equivalent to 16,000 tons of TNT. A second bomb followed three days later at Nagasaki. Over 100,000 people - most of them civilians - died instantly. Japan surrendered within days; World War II was over.

Despite the horrific display, fear in some parts of the world gradually gave way to wonder and optimism at how such a small device could produce so much energy. Even so, weapons development continued. Russia's first plutonium-producing reactor (plutonium does not occur in nature) came online at Mayak in 1948, followed by their first atomic bomb test in the deserts of Kazakhstan during August of 1949.[15] [16] Outside the Soviet Union, attention in the West turned towards applying fission's unprecedented energy potential to civilian purposes. Five days before Christmas of 1951, America's small 'Experimental Breeder Reactor 1' became the world's first electricity-producing reactor when it generated sufficient

[13] Lakoff, Andrew. *Disaster and the Politics of Intervention.* New York: Columbia University Press, 2010. Page 16.
[14] Powaski, Ronald E. *March to Armageddon: The United States and the Nuclear Arms Race, 1939 to the Present.* New York: Oxford University Press, 1987. Page 22.
[15] Cochran, Thomas B., Robert S. Norris, and Oleg Bukharin. *Making the Russian Bomb: From Stalin to Yeltsin.* Boulder, CO: Westview Press, 1995.
[16] Burr, William. *"U.S. Intelligence and the Detection of the First Soviet Nuclear Test, September 1949."* The National Security Archive, George Washington University. September 22, 2009. Accessed February 24, 2016. http://nsarchive.gwu.edu/nukevault/ebb286/.

electricity to light four 200-watt light bulbs.[17] Two years later, America's President Eisenhower announced the 'Atoms For Peace' program during a speech in which he pledged the United States', *"determination to help solve the fearful atomic dilemma - to devote its entire heart and mind to find the way by which the miraculous inventiveness of man shall not be dedicated to his death, but consecrated to his life."*[18] Part genuine attempt to push for civilian nuclear infrastructure and further research, part propaganda program to silence global critics of nuclear energy and provide a cover for a nuclear arms build-up, Atoms For Peace ultimately lead to the creation of America's nuclear power stations.[19]

One of Russia's existing military plutonium production reactors was modified for electricity generation, and in June 1952, AM-1 - short for 'Peaceful Atom 1' in Russian - became the world's first civilian nuclear power station, generating 6-Megawatts electric (MWe).[20] It was a graphite-moderated, water-cooled configuration, which served as a prototype for Chernobyl's RBMK reactors. Two years later, Queen Elizabeth II opened Britain's first commercial 50 MWe nuclear reactor at Windscale, as the government announced that Britain had become, *"the first station anywhere in the world to produce electricity from atomic energy on a full industrial scale."*[21]

Both dominant superpowers recognised the obvious potential naval benefits of a power source that only needs to be refuelled every few years, and worked hard to reduce the scale of their reactor designs. By 1954, miniaturisation had progressed

---

[17] Michal, Rick. *Fifty Years Ago in December: Atomic Reactor EBR-1 Produced First Electricity.* Report. American Nuclear Society, 2001.
[18] *"60 Years of Atoms for Peace."* Nuclear Engineering International. January 23, 2014. Accessed February 24, 2016. http://www.neimagazine.com/features/feature60-years-of-atoms-for-peace-4164653/.
[19] Ibid.
[20] Josephson, Paul R. *Red Atom: Russia's Nuclear Power Program from Stalin to Today.* New York: W.H. Freeman, 2000. Page 2.
[21] Taylor, Simon. *Privatisation and Financial Collapse in the Nuclear Industry: The Origins and Causes of the British Energy Crisis of 2002.* London: Routledge, 2007. Page 13.

far enough for the United States to launch the world's first nuclear submarine, the USS Nautilus, and both America and Russia had nuclear-powered surface ships within a further five years.

In 1973, the first high power RBMK-1000 reactor - the same type used at Chernobyl, which was under construction at the time - started up in Leningrad. America and most Western countries had now settled on a Pressurised Water Reactor design - water moderated and cooled - as the safest option. From the late 1970s until early 2000s, construction of new reactors stalled: a consequence both of the world's reaction to the Chernobyl and Three Mile Island incidents, and of improvements to the power capacity and efficiency of existing reactors. Nuclear power reached its peak in terms of number of reactors operating in 2002, with 444 in use, but it wasn't until 2006 that the highest level of nuclear-generated electricity record was set: 2,660 Terawatt-hours for the calendar year.[22]

As of 2011, nuclear power provided 11.7% of the world's electricity, with over 430 commercial nuclear reactors operating in 31 countries.[23] Combined, they generate 372,000 Megawatts of electricity. The current largest nuclear plant is Japan's Kashiwazaki-Kariwa Nuclear Power Plant, which generates 8000MW from 7 reactors, though it is not currently in use. France is the country most dependent on nuclear power, providing roughly 75% of its electricity through nuclear power plants, while America and Russia both hover around the 20% mark. Slovakia and Hungary were the only other countries to produce more than 50% of their electricity from nuclear power

---

[22] "Nuclear Power in the World Today." World Nuclear Association. January 2016. http://world-nuclear.org/information-library/current-and-future-generation/nuclear-power-in-the-world-today.aspx
[23] 2013 Key World Energy Statistics. Report. Paris: International Energy Agency, 2013.

as of the end of 2014, though Ukraine, where Chernobyl is situated, still relies on nuclear for 49% of its energy.[24]

Nuclear electricity has become the power source of choice for many large naval vessels. This peaked in the early 1990s, when there were more nuclear reactors in ships (mostly military - over 400 in submarines[25]) than there were generating electric power in commercial power plants worldwide.[26] This number has since dwindled, but there are still some 150 ships and submarines containing nuclear reactors. Russia is constructing the world's first floating nuclear power station barge for use in the Arctic, which could be towed to wherever power is needed. Containing two modified naval reactors from ice-breakers and operating at a capacity of 70MW, the *Akademik Lomonosov* is expected to be delivered in September 2016.[27] While the Russians will claim the title of first barge to produce nuclear power, floating power stations are not a new idea. The United States built the first floating nuclear power station inside a converted ex-WWII Liberty Ship in the late '60s, though none operate today. China is also entering the market, and expects its first floating nuclear power station to begin generating electricity some time in 2020.[28]

Previous Accidents

It is impossible to say for certain how many people have died as a result of nuclear accidents, because cancers and other

---

[24] *Nuclear Power Reactors in the World*. Report. Vienna: International Atomic Energy Agency, 2015.
ISBN 978–92–0–104915–5
[25] Shultis, J. K., Faw, R. E. *Fundamentals of Nuclear Science and Engineering*, Marcel Dekker, 2002, Page 340.
[26] *"Nuclear Weapons at Sea."* Bulletin of the Atomic Scientists. September 1990. Pages 48–49.
[27] *"Floating Plant To Be Delivered in 2016."* World Nuclear News. Accessed February 27, 2016.
http://www.world-nuclear-news.org/NN-floating-plant-to-be-delivered-in-2016-23101401.html.
[28] *"CGN to Build Floating Reactor."* World Nuclear News. Accessed February 27, 2016.
http://www.world-nuclear-news.org/NN-CGN-to-build-floating-reactor-1301164.html.

medical disorders caused by exposure to radiation are often indistinguishable from any other cause. Only estimates can be made. As with Marie Curie, it is likely that many of the early pioneers of radiation research (and early patients receiving overpowered X-rays[29]) were killed later in life - via cancer or radiation-related illnesses - by that which they studied. Even though Curie's work deteriorated her health - and the health of her colleagues - until her eventual death in 1934, she continued to deny the hazards of radiation. Curie's two children - who continued her work and won their own Nobel Prize - were also killed by radiation.[30] Even deaths resulting from acute radiation syndrome have no reliable statistics, as the Soviet Union covered up all serious accidents until the Chernobyl disaster. It is possible that secretive, nuclear-capable countries notorious for bureaucratic corruption such as Pakistan, Iran and North Korea may continue to do so.

There are around 70 nuclear and radiation accidents involving fatalities on public record, almost all of which resulted in less than 10 deaths,[31] although there have without doubt been more which will have been kept hidden. Interestingly, many of these events are attributed to miscalibration or theft of medical radiotherapy equipment.

For instance, more than 240 people were exposed to radiation in Goiânia, Brazil in September 1987, after thieves dismantled a steel and lead capsule stolen from a nearby semi-demolished hospital. The capsule, which contained radioactive

---

[29] Sansare, K., V. Khanna, and F. Karjodkar. "*Early Victims of X-Rays: A Tribute and Current Perception.*" Dentomaxillofacial Radiology 40, no. 2 (2011): 123-25. Accessed February 27, 2016. http://www.ncbi.nlm.nih.gov/pmc/articles/PMC3520298/pdf/dmf-40-123.pdf.
[30] Grady, Denise. "*A Glow in the Dark, and a Lesson in Scientific Peril.*" The New York Times. 1998. Accessed February 27, 2016. http://www.nytimes.com/1998/10/06/science/a-glow-in-the-dark-and-a-lesson-in-scientific-peril.html.
[31] Johnston, W. Robert, Ph.D. "*Radiation Accidents and Other Events Causing Radiation Casualties.*" Johnston Archive. January 20, 2014. Accessed February 27, 2016. http://www.johnstonsarchive.net/nuclear/radevents/radaccidents.html.

caesium from a radiotherapy machine, was stored in the back garden of one of the men. There, over the course of several days, during which both thieves became ill, the pair hacked away at the capsule until they pierced its protective steel casing. The men attributed their symptoms to something they had eaten, not suspecting their loot, and subsequently sold the compromised capsule to a scrapyard dealer named Devair Ferreira. That night, Devair noticed the material inside gave off a blue glow and assumed it to be valuable - even supernatural. To protect it, he stored the capsule in the home he shared with his wife Gabriela, and distributed powder and fragments among friends and family. This included Devair's brother, who gave some of the caesium powder to his six-year-old daughter. Enticed by the magical blue glow, she played with it, spreading it on herself like glitter, and ingested the radioactive particles. Two of Devair's employees spent several days further dismantling the capsule, to extract the lead it contained.

Gabriela was the first to notice that she and everyone around her was becoming seriously ill. Despite being told by a doctor that she, too, was having an allergic reaction to something she ate, she was convinced the culprit was the unusual material that had so fascinated her family. Gabriela reclaimed the capsule from a second scrap merchant, to whom it had now been resold, and took it - by bus - to a nearby hospital, where she declared that it was, *"killing [her] family"*.[32] Her foresight prevented the incident from being far more serious.

The caesium then sat unidentified in a courtyard until the next day, when a visiting medical physicist, who had been asked to investigate by a doctor at the hospital, *"arrived just in time to dissuade the fire brigade from their initial intention of picking up the source*

[32] Delves, D., and S. Flitton. *The Radiological Accident in Goiânia*. Publication. Vienna: International Atomic Energy Agency, 1988. ISBN 92-0-129088-8. Page 26.

and throwing it into a river".[33] Gabriela perished, along with the little girl and Devair's two employees. Devair Ferreria himself survived, despite receiving a higher dose than any of the four fatalities. Because the capsule had been opened and transported several times during the two-week incident, several areas of the city were contaminated, necessitating the demolition of multiple buildings.[34]

The total number of fatalities from accidents relating to civilian nuclear power is relatively low - far lower than deaths related to conventional coal, oil or hydro-power accidents.

To place this in perspective, consider the death tolls of the worst conventional power-related accidents. Coal mining, notorious for being dangerous, contributes a huge number of deaths. A list of just 32 notable coal mining accidents totals almost 10,000 fatalities, while all American coal-mining accidents since 1839 account for over 15,000 deaths. [35] [36] The worst of these incidents occurred on April 26th 1942, exactly 44 years prior to the Chernobyl disaster, when a gas explosion at China's Benxihu Colliery lead to the deaths of 1,549 miners.[37]

The Nigerian National Petroleum Corporation's Jesse Oil Pipeline exploded in 1998, killing over 700 people - one of dozens of similar incidents in the country. Its exact cause was never determined because everyone in the vicinity was killed, but the explosion was either due to poor maintenance or - just as

[33] Ibid. Page 27.
[34] Delves, D., and S. Flitton. *The Radiological Accident in Goiânia*. Publication. Vienna: International Atomic Energy Agency, 1988. ISBN 92-0-129088-8
[35] Wikipedia. Accessed February 27, 2016. https://en.wikipedia.org/wiki/Mining_accident. Not a reliable source in itself, but I'm using the list to illustrate a point.
[36] *"All Mining Disasters: 1839 to Present."* Centers for Disease Control and Prevention. February 26, 2013. Accessed February 27, 2016.
http://www.cdc.gov/niosh/mining/statistics/content/allminingdisasters.html.
[37] *"Honkeiko Colliery Mining Disaster | China [1942]."* Encyclopedia Britannica Online. Accessed February 27, 2016. http://www.britannica.com/event/Honkeiko-colliery-mining-disaster.

likely - deliberate sabotage by scavengers seeking to steal oil.[38] Another striking oil/gas accident happened near the Russian city of Ufa. When a leak sprang in a large gas pipeline near a remote section of the Trans-Siberian railway, instead of locating and fixing it, workers increased the pipe's gas pressure to compensate. This gradually filled the valley through which it ran with a flammable mixture of benzine and propane-butane, until people up to 5 miles away reported smelling gas. On June 4th, 1989, two trains carrying a total of around 1,200 family holidaymakers, running in opposite directions, passed each other near the leaking pipeline. Sparks from their wheels ignited the lingering gas, triggering an explosion of, *"frightening might"* - 10,000 tons of TNT. Both locomotives and their 38 carriages were incinerated and flung from the tracks, according to Mikhail Moiseyev, the Army's General Chief of the Soviet General Staff. The explosion was so powerful, *"that it felled all trees within 4 kilometers,"* he said. The accident claimed the lives of 675 people, over 100 of whom were children.[39] [40] [41]

Hydro-power's most catastrophic accident occurred during Super-Typhoon Nina in 1975, after a year's worth of rain fell on China's Henan Province in 24 hours. The Beijing-based Central Meteorological Observatory had predicted 100mm of rainfall, leaving people unprepared for what followed. At its peak, 190mm fell in a single hour.[42] *"When the rain continued, the days were*

[38] Aigbogan, Frank. *"Pipeline Explosion Kills 700."* Ludington Daily News, October 22, 1998. https://news.google.com/newspapers?id=h_hOAAAAIBAJ&sjid=f0wDAAAAIBAJ&pg=7259,4887108&dq=nigerian pipeline explosion&hl=en.
[39] Keller, Bill. *"500 on 2 Trains Reported Killed By Soviet Gas Pipeline Explosion."* The New York Times (New York), June 4, 1989. http://www.nytimes.com/1989/06/05/world/500-on-2-trains-reported-killed-by-soviet-gas-pipeline-explosion.html.
[40] *"Careless Workers Blamed For Explosion."* Observer-Reporter (Washington), June 6, 1989. Accessed February 27, 2016. https://news.google.com/newspapers?id=9M5dAAAAIBAJ&sjid=UV0NAAAAIBAJ&pg=1104,1780217&dq=propane pipeline leak&hl=en.
[41] *"Russia Remembers 1989 Ufa Train Disaster."* Sputnik News. June 4, 2009. Accessed February 28, 2016. http://sputniknews.com/russia/20090604/155167464.html.
[42] *"Hurricanes: Science and Society: 1975- Super Typhoon Nina."* Hurricane Science. Accessed March 18, 2016. http://www.hurricanescience.org/history/storms/1970s/typhoonnina/.

like nights as rain fell like arrows," survivors were quoted as saying by official records. *"The mountains were covered all over by dead sparrows after the rain."* Just after 1am on August 8th, the Banqiao Dam failed with what sounded, *"like the sky was collapsing and the Earth was cracking."*[43] An unstoppable deluge of water then prompted a chain reaction that overwhelmed 61 other dams and reservoirs. The resulting 11 kilometer-wide, 50 km/h wave ultimately killed a staggering 171,000 people, destroyed the homes of 11 million more, and wiped out entire communities.[44]

A number of nuclear accidents are worth highlighting. One early example is that of a 6.2-kilogram piece of plutonium, which went critical on two separate incidents at the Los Alamos nuclear research laboratory in New Mexico, US. It was subsequently given the nickname 'The Demon Core'. The first occasion occurred on August 21st, 1945, when Harry Daghlian, working alone, dropped a neutron-reflective brick onto the core by mistake, causing an instantaneous and uncontrolled chain reaction.[45] He knew what was happening, but had to partially disassemble his experiment to remove the brick, by which time he had already received a fatal dose. He died twenty-five days later. Despite a review of safety protocol following the accident, another event with the same piece of plutonium occurred less than a year later when physicist Louis Slotin allowed two neutron-reflecting half-spheres to envelop the core by accident, thus causing it to go critical. Leaning over the core, he received a fatal dose in less than a second and died after, *"a total*

[43] *"After 30 Years, Secrets, Lessons of China's Worst Dams Burst Accident Surface."* People's Daily Online. October 1, 2005. Accessed February 28, 2016. http://en.people.cn/200510/01/eng20051001_211892.html.
[44] *"Typhoon Nina-Banqiao Dam Failure | Chinese History [1975]."* Encyclopedia Britannica Online. June 4, 2014. Accessed February 28, 2016. http://www.britannica.com/event/Typhoon-Nina-Banqiao-dam-failure.
[45] McLaughlin, Thomas P., Shean P. Monahan, and Norman L. Pruvost. *A Review of Criticality Accidents.* Report. Oak Ridge: Los Alamos National Laboratory, 2000.

disintegration of bodily functions," nine days later.[46] Following this second accident, hands-on experiments halted and special remote control machines were used instead. After the war, scientists placed the Demon Core into a nuclear bomb and detonated it underwater at Bikini Atoll as part of America's Operation Crossroads - a study intended to test the effects of nuclear weapons on Navy ships.

Britain's worst nuclear accident was a direct consequence of the short-sighted conversion of the two existing plutonium-producing reactors at Windscale (now Sellafield) in Cumbria to instead produce tritium, which is required for a thermonuclear bomb. The graphite-moderated, air-cooled reactors were not well suited to the task, which required a hotter, more intense fission reaction than they were designed for. Engineers made modifications inside the core that enabled the production of tritium at the cost of reduced safety. When initial tests succeeded with no apparent problems, full-scale production of tritium began unabated. Nobody knew that modifying the reactor had dangerously changed the distribution of heat within the core - the reactor was now growing far too hot in areas which had earlier been cool and lacked the proper sensors for measuring temperature. When the Windscale reactors were designed and built, British scientists were inexperienced with how graphite responds to being bombarded with neutrons, and were unaware that it '*suffers dislocations in its crystalline structure, causing a build-up of potential energy,*' which could then spontaneously escape in a dangerous burst of heat. The problem was not discovered until the reactors became operational, by which time it was too late for a redesign. A solution came in the somewhat unreliable form

---

[46] "*May 21, 1946: Louis Slotin Becomes Second Victim of "Demon Core".*" American Physical Society. Accessed February 28, 2016.
https://www.aps.org/publications/apsnews/201405/physicshistory.cfm.

of a slow annealing process, where the graphite was heated and then allowed to cool, which returned it to its initial state with a gradual release of built up energy.

On October 7th, 1957, workers at Windscale performed a routine annealing process by heating up and then shutting down the reactor to wait for it to cool, but soon noticed that the release of energy was not happening as expected. The operators heated the core a second time, but by the morning of the 10th realised something was wrong - the core temperature should have fallen as the graphite energy release slowed, but it had not. Uranium fuel inside the reactor had caught fire. (Note, it was first reported to be a graphite fire, but later analysis showed it was a uranium fire.) Unaware of this critical piece of information, the operators increased the flow of air being blown into the core to help it cool down, but this only fanned the flames. At this point they noticed the radiation monitors mounted on top of the chimney were off the scale. A quick manual inspection of the reactor revealed that it was on fire, and had been for almost 2 full days. After frantic efforts to first use carbon dioxide and then water to extinguish the flames, Windscale's manager, Tom Tuohy, evacuated all but the vital personnel, shut off the cooling fans and closed the ventilators. He then climbed up the towering chimneystack several times for a direct view down into the rear of the reactor to be sure the fire was out. He later said, *'I did stand to one side, sort of hopefully, but if you're staring straight at the core of a shutdown reactor you're going to get quite a bit of radiation.'*[47]

This incident - dire as it was - would have been a disaster had it not been for "Cockroft's Folly". Sir John Cockroft was the Director of Britain's Atomic Energy Research Establishment

---

[47] Goldman, Lawrence. *Oxford Dictionary of National Biography, 2005-2008.* Oxford: Oxford University Press, 2009. Page 1137.

and had won the 1951 Nobel Prize in Physics, along with Ernest Thomas Sinton Walton, *"for their pioneer work on the transmutation of atomic nuclei by artificially accelerated atomic particles."*[48] Mid-way through Windscale's construction, Cockroft intervened and insisted that expensive radiation filters be retrofitted, overruling all objections. His filters were added, resulting in the iconic chimney bulges which came to be known as 'Cockroft's Folly' - until their existence prevented a catastrophic spread of radioactive particles across the landscape. The full facts of the accident were not made public for nearly 30 years, but a 1983 report by the National Radiological Protection Board estimated that 260 people were likely to have contracted thyroid cancer because of the incident, and over 30 more will have either already died or, *"sustained genetic damage that will bring disease or death to their descendents".*[49] The Windscale incident was regarded as the worst reactor accident until Three Mile Island, and is a fascinating story in and of itself, I recommend further reading.[50] [51] [52] [53]

America's first serious reactor accident, and the only known fatal reactor incident in US history, took place on January 3rd 1961, at the US Army's experimental SL-1 reactor. Engineers were performing maintenance requiring the large, main control rod to be disconnected from its drive motors. Reconnection

[48] "Physics Laureates: Fields." The Nobel Prize. Accessed March 18, 2016. http://www.nobelprize.org/nobel_prizes/physics/fields.html.
[49] *National Radiological Protection Board Assessment of the Radiological Impact of the Windscale Reactor Fire, October 1957: The Collective Radiation Dose Received by the Population; Addendum Concerning the Release of Polonium and Other Radionuclides.* Report. National Radiological Protection Board, 1983. http://discovery.nationalarchives.gov.uk/details/r/C11541740
[50] Arnold, Lorna. *Windscale, 1957: Anatomy of a Nuclear Accident.* New York: St. Martin's Press, 1992.
[51] Hubbell, M. W. *The Fundamentals of Nuclear Power Generation: Questions & Answers.* Bloomington, IN: AuthorHouse, 2011. Pages 100-103.
[52] Mahaffey, James A. *Atomic Awakening: A New Look at the History and Future of Nuclear Power.* New York: Pegasus Books, 2009. Chapter 3.
[53] Poole, Mike, John Dainton, and Swapan Chattopadhyay. *"Cockcroft's Subatomic Legacy: Splitting the Atom."* CERN Courier - International Journal of High-Energy Physics. November 20, 2007. Accessed February 28, 2016. http://cerncourier.com/cws/article/cern/31864.

necessitated that the operator, Army Specialist John Byrnes, manually lift the rod up by a few centimeters. He withdrew the rod too far, causing the reactor to go critical in an instant. Water inside the core explosively vaporised, causing a pressure wave to hit the lid from inside the reactor and launching the reactor vessel upwards, firing the control rods and shield plugs from their housings. One shield plug penetrated up through Construction Electrician Richard C Legg's groin and out of his shoulder, impaling and pinning him to the ceiling. He had been standing on top of the reactor. Barnes himself was killed by water and steam, and a nearby trainee died later of his injuries. Some suggest that this may not have been an accident at all, but rather a murder-suicide, as Byrnes suspected his wife was having an affair with another operator on his shift.[54]

Two submarine reactor accidents stand out. On July 4th 1961, Soviet ballistic missile submarine K-19 developed a serious leak in its reactor coolant system, causing a complete failure of the coolant pumps. Even though the control rods had been inserted into the core to neutralise the reaction, decay heat (the process of decaying radioisotopes creating heat as they lose energy - the same thing is a significant contributor to the heat at the Earth's core) rose the temperature inside to 800°C. During construction, a welder had allowed a drop of solder to land on a coolant pipe, causing a microscopic crack. During a training exercise, the crack burst open under pressure. Captain Nikolai Zateyev realised he had no choice but to create a makeshift cooling system for the reactor by cutting off an air vent valve and welding a water pipe to it. *"It would have been Chernobyl, only 30 years earlier,"* said crew member Alexander Fateyev. The

[54] Stacy, Susan M. *Proving the Principle: A History of the Idaho National Engineering and Environmental Laboratory, 1949-1999*. ID Falls, ID: Idaho Operations Office of the Dept. of Energy, 2000. Chapter 15, 16.

emergency solution worked, but the whole crew received large doses of radiation and the six brave men who entered the reactor compartment to work on the pipes died of radiation poisoning within weeks. Sixteen more would follow them. *"Right on the spot, their appearances began changing,"* recalled Captain Zateyev, after the fall of the Soviet Union. *"Skin not protected by clothing began to redden, face and hands began to swell. Dots of blood began to appear on their foreheads, under their hair. Within two hours we couldn't recognize them. People died fully conscious, in terrible pain. They couldn't speak, but they could whisper. They begged us to kill them."* The event was later depicted in the movie 'K19: The Widowmaker', starring Harrison Ford.[55] [56]

Over two decades later on August 10th, 1985, the Echo-II class submarine K-431 sat on the choppy waters of the Chazhma Bay naval facility southeast of Vladivostok, at the tri-border junction with Russia, China and North Korea. The 20-year-old sub was on the final stage of a 10-step refuelling process. This required the 12-ton reactor lid to be detached from its control rods, then lifted by a crane arm extended across the water from a nearby refuelling service ship, to allow new fuel assemblies to be placed. The reactor lid had been replaced, the control rods reattached and the cooling system refilled with water, but workers on the submarine discovered that the lid had not formed a perfect seal. Without seeking the proper authorisation, they craned-up the lid by a few centimeters to fix the problem, leaving the rods attached to save time. At this worst possible moment, a Navy torpedo boat sped by, creating a wake violent enough to rock the refuelling ship and its crane

[55] Bivens, Matt. *"Horror of Soviet Nuclear Sub's '61 Tragedy Told."* The Los Angeles Times (Los Angeles), January 3, 1994. Accessed February 28, 2016. http://articles.latimes.com/1994-01-03/news/mn-8123_1_soviet-nuclear-submarine.
[56] Dolgodvorov, Vladimir. *"СУБМАРИНА, СБЕРЕГШАЯ МИР."* Газета Труд. November 21, 2002. Accessed March 03, 2016. http://www.trud.ru/article/21-11-2002/49488_submarina_sberegshaja_mir.html.

arm. The attached lid and control rods lurched away from the core and the reactor instantly went critical, causing a steam explosion that blew the core's contents out of the compartment and destroyed the submarine's pressure hull. Eight officers and two workers were killed by the explosion, while an additional 290 workers received significant doses of radiation in the 4-hour battle to bring the resulting fire under control.[57] The accident remained secret until a book of declassified documents was published in 1993, following the fall of the Soviet Union.

Kyshtym

The event that came to be known as the Kyshtym Disaster happened near Russia's closed city of Chelyabinsk-65, 120 kilometers from the border with Kazakhstan. The existence of closed cities was a well-guarded secret during the Cold War - even among the USSR's own citizens - because they housed workers of nearby nuclear facilities, weapons factories and other significant industrial sites. They did not appear on any map or road sign, visitors were prohibited without express permission from the Government, and residents who left the city were forbidden from discussing where they lived or worked with outsiders. As a result of this secrecy, the disaster was named after Kyshtym, the nearest known town. Besides being the location of one of Russia's largest tank factories, Chelyabinsk-65 was near to the Mayak nuclear plutonium-producing reactors (for nuclear weapons) and reprocessing plant - one of the country's biggest nuclear facilities, and the site where their first nuclear weapon was produced. The Soviet Government was not

---

[57] Takano, Makoto, Vanya Romanova, Hiromi Yamazawa, Yuri Sivintsev, Keith Compton, Vladimir Novikov, and Frank Parker. "*Reactivity Accident of Nuclear Submarine near Vladivostok.*" Journal of Nuclear Science and Technology 38, no. 2 (October 16, 2000): Pages 143-57. http://www.tandfonline.com/doi/abs/10.1080/18811248.2001.9715017.

known for its compassion for the safety its people or the environment, and Mayak was no exception, as the site played host to a long list of nuclear accidents and biological atrocities in the decades after its completion in 1948. By the time of the catastrophe which would claim Kyshtym's name, the Mayak facility had already contaminated the surrounding area with constant dumping of nuclear and chemical waste into the nearby Techa-Iset-Tobol river system and lakes, to such an extent that it would be regarded as the most contaminated place on Earth decades later.

Mayak cooled some of its nuclear waste in buried steel and concrete storage tanks, each containing $300m^3$ (around 80 tons) of materials. At some time during September 1957, one of the tanks' cooling systems failed. Nobody noticed as the temperature within began to rise due to decay heat, even as it reached a temperature of approximately 350°C. On the afternoon of September 29th, 1957, built-up pressure caused the tank to explode with the force of 70 - 100 tons of TNT, throwing off the 160-ton concrete lid, damaging the two adjacent tanks and spewing 740,000 terabecquerels of radioactive particles into the air - twice the amount released by Chernobyl.

The prevailing northeasterly wind carried the radioactive plume over an area of up to 20,000 square kilometers (km^2), with serious contamination covering around $800km^2$. Reliable statistical health information is impossible to find as officials hid the accident from public view and no registry was created to track the health of those affected. After an initial (unjustifiable) delay of a week, over 10,000 people were evacuated from their homes during the following two years. Doctors diagnosed those who fell ill with 'the special disease', because they could not refer to radiation as long as the Mayak facility was a secret. It worked: the accident remained hidden until 1976 when Zhores

Medvedev (who went on to write the excellent 'Legacy of Chernobyl') exposed the event in an article for New Scientist. The event was then given a rating of 6 on the International Nuclear Events Scale, making it the third worst nuclear accident in history. Lev Tumerman, a Soviet scientist who had passed through the area in 1960, supported Medvedev's assertions, stating that, *"about 100 kilometres from Sverdlovsk, a highway sign warned drivers not to stop for the next 20 or 30 kilometers and to drive through at maximum speed. On both sides of the road, as far as one could see, the land was 'dead': no villages, no towns, only the chimneys of destroyed houses, no cultivated fields or pastures, no herds, no people...nothing."*[58] It transpired that the CIA had known about the incident for over fifteen years, but had kept silent because they didn't want to spread fear of the United States' own nuclear facilities among the population.

Mayak was the location of another serious radiation accident ten years later. Lake Karachay is a small lake on the site which had been used as a dump for radioactive waste for over a decade. Dumping into the lake continued after the explosion and by the mid-1960s it was so contaminated that standing on its shores at the time would give you a lethal dose within an hour. 1965 and '66 were particularly dry years, causing the lake to begin to dry out. During a drought in the spring of 1967, low-level areas of the lake evaporated completely, exposing radioactive sediments to the atmosphere. A violent storm swept through the area, blowing the contaminated particles several hundred kilometers from the almost bone-dry lake bed and depositing 185,000 terabecquerels (the same amount released by the Hiroshima bomb) of radioactivity onto half a million people - the same people irradiated by the Mayak explosion ten years

[58] Soran, Diane M., and Danny B. Stellman. *An Analysis of the Alleged Kyshtym Disaster*. Report. New Mexico: Los Alamos National Laboratory, 1982.

earlier. Years later, the lake was filled with thousands of hollow concrete blocks to prevent the same thing from ever happening again. [59] [60] [61] [62] [63]

Soviet accidents were not isolated to military installations. Operators at the Beloyarsk Nuclear Power Plant received serious radiation exposure in 1977 after a partial meltdown, and again a year later during a reactor fire. Despite all these events, Soviet authorities continued to maintain in public that their nuclear program was absolutely safe. Lev Feoktistov, Deputy Director of the I. V. Kurchatov Institute of Atomic Energy - now Russia's leading nuclear research and development institution, named after its founder - co-authored an article in Soviet Life magazine a year before the Chernobyl accident. In it, he stated that, *"in the 30 years since the first Soviet nuclear power plant opened, there was not a single instance when plant personnel or nearby residents have been seriously threatened: not a single disruption in normal operation occurred that would have resulted in the contamination of the air, water or soil. Thorough studies conducted in the Soviet Union have proved completely that nuclear power plants do not affect the health of the population."*[64]

Three Mile Island

The most well-known accident at a nuclear plant prior to Chernobyl occurred at the Three Mile Island (TMI) power station in Pennsylvania, United States on March 28th, 1979, when

[59] Feshbach, Murray. *Ecological Disaster: Cleaning Up the Hidden Legacy of the Soviet Regime.* New York: Twentieth Century Fund Press, 1995.
[60] *"Production Association 'MAYAK'"* Global Security. Accessed February 28, 2016. http://www.globalsecurity.org/wmd/world/russia/chelyabinsk-65_nuc.htm.
[61] Postol, T. A. *"The Incident in Chelyabinsk."* Science 206, no. 4416 (October 19, 1979): 326-27. http://science.sciencemag.org/content/206/4416/326.
[62] Rabl, Thomas. *"The Nuclear Disaster of Kyshtym 1957 and the Politics of the Cold War."* Environment & Society Portal, Arcadia 2012, no. 20. Rachel Carson Center for Environment and Society. http://www.environmentandsociety.org/node/4967.
[63] *"Ural Mountains Radiation Pollution."* American University, Washington DC. Accessed February 29, 2016. http://www1.american.edu/projects/mandala/TED/ural.htm.
[64] Hopkins, Arthur T. *Unchained Reactions: Chernobyl, Glasnost, and Nuclear Deterrence.* Washington, DC: National Defense University Press, 1993. p13.

a cooling fault lead to the meltdown of the site's brand-new second reactor. Although nobody was injured, it is considered to be the worst accident in the history of US nuclear energy. Much like Chernobyl, it combined a complicated series of oversights and mistakes to create a near-disaster.

Eleven hours before the accident began, while attempting to clean a condensate filter, a stubborn blockage compelled workers to blow compressed air into a water pipe, intending to let the water's force clean the filter. This worked, but it also caused an inadvertent trickle of water to leak into the feedwater pumps' control system. The resulting fault went undiscovered until the accident concluded.

Eleven hours later, at 4am, a minor malfunction in the secondary, non-nuclear water cooling circuit prevented proper heat dissipation and caused the primary coolant temperature to rise. TMI's reactor shut itself down, halting the chain reaction, but decay heat continued to raise the core temperature. This alone wasn't a problem, as nuclear reactors are designed with decay heat in mind and multiple automatic, redundant, independent safety systems are in place to prevent an accident. By an unlucky coincidence, however, the three auxiliary water coolant pumps that also activated could not pump any water because their valves were closed for routine maintenance. Decay heat in the core created a pressure build-up much like it had at Mayak, prompting the pressuriser's pilot-operated relief valve (PORV) to open, which stabilised the pressure level. Then things started to go wrong. The mechanical fault from 11 hours prior came into play, preventing the valve from closing again. Reactor 2's operators incorrectly assumed that the valve had closed, because their control panels indicated that a 'close' signal had been sent to it - not what the valve's actual position was. As a result, they failed to notice that coolant was escaping from the system for several hours, leading them to make several wrong-moves.

With coolant rapidly escaping, the control computer injected emergency water from pressurised tanks into the system to compensate. Although a significant volume of this injected water also escaped through the PORV, enough was flushed past the pressuriser's water sensors to trick the operators into believing that there was actually *too much* water in the cooling system. They responded by reducing the flow of replacement water, unintentionally starving the reactor of water and allowing dangerous steam to build up within the primary cooling system. When steam bubbles form in fluid and subsequently collapse, they emit high-pressure shock waves that can damage the pipes. This is known as cavitation. TMI's control room personnel, who were still under the impression there was sufficient water travelling around the cooling system, turned off the pumps to prevent this. Diminishing water levels gradually exposed the top of the fuel elements inside the core, allowing them to reach extreme temperatures and melt, which released radioactive particles into the remaining water. During all of this, the reactor operators struggled to figure out what was wrong.

It was only when the control room shift changed at 6am that fresh eyes noticed the PORV temperature was higher than expected. At 6:22 am operators closed a backup block valve between the relief valve and the pressuriser. The coolant loss halted, but by now superheated steam was preventing the inertia circulation of water, so they slowly increased the pressure by injecting pressurised water into the cooling system. Over 16 hours after the disaster began, the pressure climbed high enough to restart the primary pumps without fear of cavitation. It worked: the reactor temperature fell, but not before about half of the core and 90% of the fuel's safety cladding had melted.

The event was saved from being catastrophically worse by the reactor's pressure vessel - an enormous metal shield surrounding the core, containing its molten radioactive remains. The same vital containment that Chernobyl's RBMK reactors lacked.[65] [66]

As at Chernobyl, operator error was shouted loud as the fundamental cause of the accident, but US President Jimmy Carter's own President's Commission came to more pragmatic conclusions seven months later. Their report noted many areas where improvements could be made. *"While training may have been adequate for the operation of a plant under normal circumstances, insufficient attention was paid to possible serious accidents."* It also acknowledged that some, *"operating procedures, which were applicable to this accident, are at least very confusing and could be read in such a way as to lead the operators to take the incorrect actions they did."* Problems with the confusing control interface were addressed too: *"The control room, through which the operation of the* [reactor] *is carried out, is lacking in many ways. The control panel is huge, with hundreds of alarms, and there are some key indicators placed in locations where the operators cannot see them... During the first few minutes of the accident, more than 100 alarms went off, and there was no system for suppressing the unimportant signals so that operators could concentrate on the significant alarms."* Finally, the timeless problem of failure to learn from past mistakes also contributed, as it transpired a similar incident happened at another U.S. plant over a year earlier, but American reactor operators were not informed.[67]

While these events are disturbing when taken in isolation, it's important to remember that nuclear power remains by far

[65] Walker, J. Samuel. *Three Mile Island: A Nuclear Crisis in Historical Perspective*. Berkeley: University of California Press, 2004.
[66] Rogovin, Mitchell, Director. Three Mile Island - A Report to the Commissioners and to the Public. Report. Vol. 1. Washington D.C.: Nuclear Regulatory Commission Special Inquiry Group, 1980.
[67] Kemeny, John G., Chairman. *Report of The President's Commission on the Accident at Three Mile Island*. Report. Washington D.C.: U.S. Government Printing Office, 1979.

the least harmful method of energy production overall. Using historical production data, NASA scientists calculated in 2013 that nuclear power has actually prevented an average of 1.84 million air pollution-related deaths and 64 gigatonnes of CO_2-equivalent greenhouse gas emissions that would have resulted from fossil fuel burning between 1971 and 2009.[68] That data was based on European and US plants, which tend to be cleaner than elsewhere, meaning those numbers are likely to be far higher in reality. A study by Tsinghua University associate professor Teng Fei estimates that Chinese coal pollution caused a distressing 670,000 deaths in 2012,[69] while the global average coal deaths is 170 per Terawatt-hour (TWh) of generated electricity. For comparison, data from 2012 shows that oil-generated electricity causes 36 deaths/TWh; biofuel, 24 deaths/TWh; wind power, 0.15 deaths/TWh; hydro-electricity, if you factor in the Banqiao disaster, causes 1.4 deaths/TWh, and still causes widespread devastation to the surrounding landscape if you don't. Nuclear power, including Chernobyl and Fukushima, is responsible for 0.09 deaths per Terawatt-hour.[70]

[68] Kharecha, Pushker A., and James E. Hansen. "*Prevented Mortality and Greenhouse Gas Emissions from Historical and Projected Nuclear Power.*" Environmental Science & Technology 47, no. 9 (2013): 4889-895. http://pubs.giss.nasa.gov/abs/kh05000e.html.
[69] Smith, Geoffrey. "*The Cost of China's Dependence on Coal – 670,000 Deaths a Year.*" Fortune. November 05, 2014. Accessed February 29, 2016. http://fortune.com/2014/11/05/the-cost-of-chinas-dependence-on-coal-670000-deaths-a-year/.
[70] Conca, James. "*How Deadly Is Your Kilowatt? We Rank The Killer Energy Sources.*" Forbes. June 10, 2012. Accessed February 29, 2016. http://www.forbes.com/sites/jamesconca/2012/06/10/energys-deathprint-a-price-always-paid/.

CHAPTER TWO

Chernobyl

The Chernobyl Nuclear Power Plant, officially known as the V. I. Lenin Atomic Power Plant during the Soviet era, began construction in 1970 in a remote region near Ukraine's swamp-filled northern border, 15 kilometers northwest of the small town of Chernobyl. The desolate location was chosen because of its relative proximity to yet safe distance from Ukraine's capital, a ready water supply - the River Pripyat - and the existing railway line running from Ovruc in the west to Chernigov in the east. It was the first nuclear power station ever to be built in the country, and was considered to be the best and most reliable of the Soviet Union's nuclear facilities.[71] Concurrent to the construction of the power station, the Soviet Union's ninth Atomograd - Russian for 'atomic city' - named Pripyat was being erected 3 kilometers away, for the express purpose of housing the ambitious station's 50,000 operators, builders, support staff

[71] Gelino, Nathan, Marta-Rey Babarro, Mark A. Siegler, Deepti Sood, and Craig Verlinden. *Chernobyl - Nuclear Disaster. An Accident Investigation Report Submitted for IOE491, Human Error & Complex System Failures*. Report. 2005.

and their families. Pripyat was one of the 'youngest' cities in the Soviet Union, with an average age of only 26.

To oversee the titanic operation, 35-year-old turbine expert and loyal communist Viktor Bryukhanov was plucked from his position as Deputy Chief Engineer at the Slavyanskaya thermal power plant in eastern Ukraine, and appointed as Chernobyl's Director.[72] It seems that he was genuinely liked and respected as a Director, with one of the plant's original Deputy Chief Engineers commenting, *"He is a great engineer. I really mean it."*[73] In his new capacity, Bryukhanov was responsible for overseeing construction of both the plant and city, and organising everything from the recruitment of workers to the procurement of machinery and masonry. Bryukhanov worked hard but, despite his earnest efforts, the construction suffered a plethora of problems typical of the Communist system. Thousands of tons of reinforced concrete were missing from orders, and specialist equipment was either impossible to source or of poor quality when it eventually arrived, forcing him to order the manufacture of replacements in makeshift on-site workshops.[74] Although these complications put the plant two years behind schedule, the first reactor - Unit 1 - was commissioned on the 26th of November 1977, following months of tests. Three more reactors followed: Unit 2 in 1978, Unit 3 in 1981, and Unit 4 in 1983.

All four reactors were the relatively new, Soviet-designed RBMK-1000 ('Reaktor Bolshoy Moshchnosti Kanalnyy', or 'High Power, Channel-type Reactor' in English), which output 1000 Megawatts of electrical power via two 500MW steam

[72] Medvedev, Grigoriï, and Andreï Sakharov. *The Truth About Chernobyl*. New York: BasicBooks, 1991. Page 42.
[73] Karpan, N. V. *Trial at Chernobyl Disaster*. Report. Kiev, 2001.
[74] Read, Piers Paul. *Ablaze: The Story of Chernobyl*. London: Secker & Warburg, 1993. Pages 39-40.

turbogenerators. The RBMK-1000 is a graphite-moderated, boiling water-cooled reactor; an unusual and slightly outdated combination that was designed in the 1960s to be powerful, quick, cheap and easy to build, relatively simple to maintain, and to have a long service life. Each reactor measures a massive 7 meters tall by 11.8 meters wide.[75] In 1986, fourteen of this type were in service, while another eight were under construction. Two of these were being built at Chernobyl on the night of the accident in 1986, with Unit 5 expected to be completed later that year. The four existing reactors together provided 10% of Ukraine's electricity at the time and, had Units 5 and 6 been completed, Chernobyl would have been the highest capacity, non-hydro power station in the world.[76] For reference, the world's largest hydroelectric power station by installed capacity is the Three Gorges Dam in China, which is rated for an incredible 22,500MW.[77]

Nuclear reactors use a process called nuclear fission - sometimes called 'splitting the atom' - to generate electricity. All matter is composed of atoms, and each atom is mostly empty space, with a tiny centre of protons and neutrons joined together to form a nucleus, which gives an atom most of its weight. Much of the leftover space within an atom is occupied by electrons orbiting the nucleus in the middle. The differences between atoms come from the differing number of protons and neutrons in a given nucleus. For example, the element gold contains 79 protons, and is famous for being heavy. Copper has just 29 protons, and is far less dense than gold. Oxygen only has 8

---

[75] International Safety Advisory Group. *The Chernobyl Accident: Updating of INSAG-1: INSAG-7.* Vienna: International Atomic Energy Agency, 1992.
[76] Pedraza, Jorge Morales. *Electrical Energy Generation in Europe: The Current Situation and Perspectives in the Use of Renewable Energy Sources and Nuclear Power for Regional Electricity Generation.* Berlin: Springer, 2015. Page 526.
[77] Xuequan, Mu. *"Three Gorges Breaks World Record for Hydropower Generation."* Xinhua. January 1, 2015. Accessed February 29, 2016. http://news.xinhuanet.com/english/china/2015-01/01/c_127352471.htm.

protons. Every atom will have the same number of orbiting electrons as it does protons, but atoms of the same element can have different numbers of neutrons. These different versions of the same element are known as isotopes. You could think of isotopes as being like a car with optional upgrades. Mercedes has many cars - the elements - in their lineup, but these cars may have optional extras to add: a more powerful engine; different upholstery; an expensive paint job. The car remains the same vehicle, but is now in a different form. Stable isotopes - that is, isotopes which do not undergo spontaneous radioactive decay - are called stable nuclides, while unstable isotopes are collectively known as radionuclides. Together, these two groups that resulted from fission are known as 'fission products', almost all of which are the unstable radionuclide variety. These radionuclides are waste products of the reaction and are hot and highly toxic.

The RBMK, like almost all commercial nuclear reactors, uses uranium - which has 92 protons, making it the heaviest naturally occurring element - as a fuel source. Uranium contains a mere 0.7% of the fissionable isotope uranium^{235} (92 protons and 143 neutrons), and the 190 tons of fuel in a second-generation RBMK reactor like Chernobyl's Unit 4 consists of cheap and only slightly enriched 98% uranium^{238} and 2% uranium^{235}, contained within 1,661 vertical pressure tubes. During the nuclear reaction inside a reactor core, neutrons collide with the nuclei of another uranium^{235} atom, splitting it and creating energy in the form of heat. This atomic split creates an additional two or three neutrons. These new neutrons will then collide with more U-235 fuel, splitting another uranium atom to form yet more neutrons, and so on. This process is called a fission chain reaction, and it is this reaction which creates the heat in a nuclear reactor. At the same time, additional new elements in the form of hot fission products are created.[78]

Nuclear power harnesses the same atomic reaction as a nuclear bomb, but is designed to ensure that it is physically incapable of causing a nuclear explosion, and instead controls the release of neutrons to generate the required heat. While a power station's reactor contains barely-enriched uranium or plutonium fuel, dispersed over a large area and surrounded by control rods to restrain the reaction, a nuclear bomb is designed with the specific intention of causing this same reaction to occur instantaneously and with far greater intensity, by using explosives to force two hemispheres of 90%+ enriched uranium or plutonium together.

Preventing a radioactive release is the highest priority at any nuclear facility, so power stations are built and operated with a safety philosophy of 'defense in depth'. Defense in depth aims to avoid accidents by embracing a safety culture, but also accepts that mechanical (and human) failures are inevitable. Any possible problem - however unlucky - is then anticipated and factored into the design with multiple redundancies. The goal, therefore, is to provide depth to the safety systems; akin to the way Russian dolls have several layers before reaching the core doll. When one element fails, there is another, and another, and another that still functions. The first barrier are the fuel ceramic pellets themselves, followed by each fuel rod's zirconium alloy cladding. In an ordinary modern commercial nuclear plant, the nuclear core where the fission reaction takes place would be contained inside a third barrier: an almost unbreakable metal shield enveloping the reactor, called a 'pressure vessel'. The RBMK forgoes a conventional pressure vessel and instead only

[78] Medvedev, Zhores A. *The Legacy of Chernobyl*. Oxford: Basil Blackwell, 1990. Page 245.

uses reinforced concrete around the sides of the reactor, with a heavy metal plate at the top and bottom. Adding a proper pressure vessel, built to the standards and complexity required by the RBMK design, was estimated to double the cost of each reactor. The fourth and final barrier is an airtight containment building. It is well known that nuclear containment buildings are very, very heavily reinforced, with concrete and/or steel walls often several meters thick. They are built to withstand the external impact of an airliner crashing into them at hundreds of miles-per-hour, but their other purpose is to contain the unthinkable breach of a pressure vessel. Unbelievably, the RBMK's accompanying reactor building is insufficient to be labelled as a true containment building, presumably as part of further cost saving measures.[79]

The RBMK's stunning dual lack of the most crucial containment barriers is a glaring omission that should never have been considered, let alone designed, approved and built. Select Soviet Ministers were made aware of these inadequacies before the reactors were chosen, but still the RBMK design was selected over the competing 'Vodo-Vodyanoi Energetichesky Reaktor' (VVER, or 'Water-Water Power Reactor'), a pressurised water reactor which was safer, but more expensive and marginally less powerful. Conventional wisdom at the time was that the RBMK could never cause a large-scale accident, because industry safety regulations would always be adhered to. Extra safety measures, they decided, were unnecessary.[80]

A fission reaction is enabled by what is known as a neutron moderator, which, in an RBMK reactor, is comprised of vertical graphite blocks surrounding the fuel channels. Each RBMK

---

[79] Domaratzki, Z., Chairman. *Defence in Depth in Nuclear Safety: A Report by the International Nuclear Safety Advisory Group*. Vienna: International Atomic Energy Agency, 1996.
[80] Medvedev, Grigoriĭ. *Chernobyl Notebook*. 1989. Chapter 2.

contains 1850 tons of graphite. This graphite slows - moderates - the speed of neutrons moving in the fuel, because slowed neutrons are far more likely collide with uranium^{235} nuclei and split. When playing golf, for example, if your ball is a few centimeters from the hole, you don't hit it as hard as you possibly can, you give it a slow tap to the target. It's the same principle with neutrons in a reactor. The more often the resulting atomic split occurs, the more the chain reaction sustains itself and the more energy is produced. In other words, the graphite moderator creates the right environment for a chain reaction. Think of it as oxygen in a conventional fire: even with all the fuel in the world, there will be no flame without oxygen.

Using graphite as a moderator can be highly dangerous, as it means that the nuclear reaction will continue - or even increase - in the absence of cooling water or the presence of steam pockets (called 'voids'). This is known as a positive void coefficient and its presence in a reactor is indicative of very poor design. Graphite moderated reactors were used in the USA in the 1950s for research and plutonium production, but the Americans soon realised their safety disadvantages. Almost all western nuclear plants now use either Pressurised Water Reactors (PWRs) or Boiling Water Reactors (BWRs), which both use water as a moderator and coolant. In these designs, the water that is pumped into the reactor as coolant is the same water that is enabling the chain reaction as a moderator. Thus, if the water supply is stopped, fission will cease because the chain reaction cannot be sustained; a much safer design. Few commercial reactor designs still use a graphite moderator. Other than the RBMK and its derivative, the EGP-6, Britain's Advanced Gas-Cooled Reactor (AGR) design is the only other graphite-moderated reactor in current use. The AGR will soon be joined by a new type of experimental reactor at China's Shidao Bay

Nuclear Power Plant, which is currently under construction. The plant will house two graphite-moderated 'High Temperature Reactor-Pebble-bed Modules' reactors, which are expected to begin operation in 2017.

Because of the extreme heat fission generates, the reactor core must be kept cool at all costs. This is particularly important with an RBMK, which operates at an, "astonishingly high temperature," relative to other reactor types, of 500°C with hotspots of up to 700°C, according to British nuclear expert Dr. Eric Voice. A typical PWR has an operating temperature of about 275°C. A few different kinds of coolant are used in different reactors, from gas to air to liquid metal to salt, but Chernobyl's uses the same as most other reactors: light water, meaning it is just regular water. The plant was originally going to be fitted with gas-cooled reactors, but this was eventually changed because of a shortage of the necessary equipment.[81] Water is pumped into the bottom of the reactor at high pressure (1000psi, or 65 atmospheres), where it boils and passes up, out of the reactor and through a condensator that separates steam from water. All remaining water is pushed through another pump and fed back into the reactor. The steam, meanwhile, enters a steam turbine, which turns and generates electricity. Each RBMK reactor produces 5,800 tons of steam per hour.[82] Having passed through this turbogenerator, the steam is condensed back into water and fed back to the pumps, where it begins its cycle again.

There's one major shortcoming inherent to using this method of cooling. Unlike in a typical PWR, the water entering the reactor is the same water that passes through the cooling

---

[81] International Safety Advisory Group. *The Chernobyl Accident: Updating of INSAG-1: INSAG-7*. Vienna: International Atomic Energy Agency, 1992. p33.
[82] Dubrovsky (ed.), *Construction of Nuclear Power Plants*. Page 37.

pumps and then as steam through the turbines, meaning highly irradiated water is present in all areas of the system. A PWR uses a heat exchanger to pass heat from the reactor water to clean, lower pressure water, allowing the turbines to remain free of contamination. This is better for safety, maintenance and disposal. A second problem is that steam is allowed to form in the core, making dangerous steam voids more likely, and further increasing the chances of a positive void coefficient. In ordinary boiling water reactors, which use water as both a coolant and moderator like in a PWR, this would not be such a problem, but it is in a graphite-moderated BWR.

To control the release of energy by a nuclear reactor, 'control rods' are used. RBMK control rods are long, thin cylinders, composed mostly of neutron-absorbing boron carbide to hinder the reaction. The tips of each rod are made of graphite to prevent cooling water (which is also a neutron absorber) from entering the space the rod's boron had occupied as it is withdrawn from the core, in order for that section to have a greater impact upon reactivity when reinserted. Chernobyl's 211 control rods descend down into the core from above as necessary, and are aided in their role by an extra 24 special shortened 'absorber rods'. These absorber rods ensure an even distribution of power across the entire width of the core by inserting upwards from below. The more control rods that are inserted into the reactor core, and the further they penetrate, the lower the levels of power will be. Conversely, fewer rods equals more power. Every control rod can be inserted together, penetrating as near or as far as the operator wishes, or they can be disconnected and inserted in groups, depending on requirements.[83] The RBMK control rods are incredibly slow by

---

[83] Medvedev, Grigoriï. *Chernobyl Notebook*. Moscow: Novy Mir, 1989. Chapter 2.

Western standards, taking 18-21 seconds to fully insert from their uppermost position. Some, like Canada's CANDU reactor, can take as little as 1 second.[84]

It is not well known that there was a severe accident at Chernobyl before the disaster of 1986, which resulted in the partial core meltdown of Unit 1. The incident occurred on September 9th, 1982, and remained secret for several years afterwards. Detailed and reliable reports are difficult to come by (especially in English), but it seems a coolant water control valve was closed, leading to overheating of a water channel and partial damage of the fuel assembly and graphite inside the reactor. A classified KGB report from the next day stated: *"In connection with the planned overhaul of the 1st fuel unit of the Chernobyl nuclear power plant, which is scheduled to be completed on 13 September, 1982, a trial run of the reactor was performed on 9 September 1982. When its power was increased to 20%, there was a break in one of the 1640 pressure channels/loaded fuel assemblies. At the same time, the column where the fuel assemblies are located broke. In addition, the graphite stack became partially wet."*[85] This resulted in fuel and graphite being washed out through the pipes and fission products being vented from the chimney, which in turn prevented coolant from properly reaching the reactor, leading to the partial meltdown.

Operators were unsure of what was happening for a long time and ignored warning alarms for almost half an hour. Then the KGB accident investigation seemed to ignore negligent actions of plant staff (stopping the flow of coolant on purpose). The findings of two separate organisations measuring radioactive contamination outwith the plant wildly differed, too, with a government nuclear industry commission finding almost

---

[84] International Safety Advisory Group. The Chernobyl Accident: Updating of INSAG-1: INSAG-7. Vienna: International Atomic Energy Agency, 1992. Page 4.
[85] Tykhyy, Volodymyr. *From Archives of VUChK-GPU-NKVD-KGB, Chernobyl Tragedy in Documents and Materials*. Security Service of Ukraine, 2001. Document no. 9.

no contamination at all, while a team of biophysicists from the Institute of Nuclear Research from Ukraine's Academy of Science found radiation *"hundreds of times higher than permissible levels"*.[86] Two senior figures, who would later analyse the 1986 disaster, did not agree with the official description of events either. For their part, the reactor operators on duty that day denied any wrongdoing. *"As an eyewitness of this accident and one of those involved in elimination of its consequences, I don't have much to add* [to] *the version of NIKIET* [the Scientific Research and Design Institute of Power and Technology] *that blamed the Chernobyl ATS engineer for stopping completely* [the] *water supply into the* [reactor, except that it] *has never grown into anything else but a version,"* writes Nikolai V Karpan, a senior engineer who worked at Chernobyl from 1979 to 1989. *"Both the foreman and the whole team of servicemen that carried out flow rate adjustments that day have been repeatedly denying the error inflicted upon them. On that day they worked in the usual way, in strict compliance with the regulations, according to which that a guide plate was to be installed on the regulator that would mechanically prevent complete stopping of water supply into the channel."*[87] It is likely that a flaw in the reactor design or - more probable - poor manufacturing quality was identified as a principal cause of the accident, but the politicians chose to go with the easy option and blame an operating engineer instead. One instance of human error is more palatable than acknowledging that your brand new nuclear reactor, developed and built at enormous expense, and already operating at two other existing plants, has a flaw in its design. This unofficial version of events was supported by the plant's Research Supervisor, who conducted an investigation of his own and reported: *"It turned out that the zirconium channel pipes were*

[86] Tykhyy, Volodymyr. *From Archives of VUChK-GPU-NKVD-KGB, Chernobyl Tragedy in Documents and Materials (Summary)*. Security Service of Ukraine, 2001. Page 5.
[87] Karpan, Nikolaii V. *Chernobyl. Vengeance of Peaceful Atom*. Dnepropetrovsk, 2006. ISBN 966-8135-21-0. Pages 299-300.

destroyed due to residual internal stress in their walls. The manufacturing plant had, on its own initiative, changed the production process of channel pipes, and this 'novelty' resulted in the accident in the reactor."[88] [89]

Even before the Chernobyl incident of 1982, there was another serious accident involving the RBMK design at the Leningrad Nuclear Power Plant in November 1975, when its Unit 1 suffered a partial meltdown. Detailed information is more challenging to find than the 1982 Chernobyl accident, but Viktor M. Dmitriev, a Russian nuclear engineer from the Institute of Nuclear Power Operations in Moscow, has a web page explaining what happened. The accident bears some remarkable similarities to Chernobyl's 1986 disaster. Leningrad's Unit 1 was restarting after routine maintenance and had reached 800MW when operators disconnected one of its two turbines due to a fault. To keep the reactor stable, power was reduced to 500MW and then the evening shift handed over the reins to the night shift. At 2am, someone in the control room disconnected the only remaining turbine by accident, tripping the emergency computer system and automatically shutting down the reactor. Reactor poisoning began (I'll explain this in more detail later), leaving the operators with a choice of battling the reactor back to full power or allowing it to shut down, but there would be repercussions for allowing it to happen at all. They chose - just as with Chernobyl - to raise the power. It didn't go well. *"During rising to power after shutdown, without any operator's actions to change reactivity (without lifting any rods) the reactor would suddenly reduce acceleration time by itself, i.e., inadvertently accelerate; in other words, it would try to explode,"* says V. I. Boretz, a trainee from Chernobyl who happened to be on this shift. *"The reactor acceleration was*

---

[88] Ibid.
[89] Dmitriev, Viktor Markovic, Ph.D. "*The 1982 Accident at the 1st Power Unit of ChNPP.*" The Causes of the Chernobyl Accident Are Known. Accessed March 02, 2016. http://accidont.ru/ENG/accid82.html.

stopped twice by the emergency protection system [in fact, the emergency protection was triggered more than twice, both on excess of power and on speed of its growth - Viktor M. Dmitriev]. Attempts of the operator to reduce capacity growth velocity by standard methods, lowering at the same time a group of manually controlled rods, plus four automatically controlled ones, failed, and rising to power was increasing. It was only stopped by triggering the emergency protection system." The reactor eventually reached a power of 1,720MW - almost twice its rated capacity - before it was brought under control.[90]

A government commission into the accident found serious faults with the design, and in 1976 recommended that the void coefficient be lowered, the control rod design be altered, and for 'fast-acting emergency protection' to be installed. New designs were drawn up for the rods, but were never installed on any reactors. On October 16th, 1981, a report was submitted to the KGB highlighting several concerns over the quality of construction and equipment at Chernobyl. It stated that there had been 29 emergency shutdowns during the plant's first 4 years of operation - 8 of which were caused by personnel errors, the rest from technical faults - and that, *"control equipment does not meet the requirements for reliability."* These faults had been brought to the attention of the Ministry of Power and Electrification and the design institute responsible for the reactor, *"several times,"* by the date of the report, according to the KGB, yet nothing had been done.[91]

In late 1983, Lithuania's brand new Ignalina Power Station began commission testing its first RBMK reactor and soon encountered a problem: control rods entering the reactor

---

[90] Dmitriev, Viktor Markovic, Ph.D. *"Accident at the Leningrad NPP (LNPP) in 1975."* The Causes of the Chernobyl Accident Are Known. Accessed March 02, 2016. http://accidont.ru/ENG/LAES.html.
[91] Tykhyy, Volodymyr. *From Archives of VUChK-GPU-NKVD-KGB, Chernobyl Tragedy in Documents and Materials.* Security Service of Ukraine, 2001. Document 8.

together caused a power surge. This is basically what caused the Chernobyl disaster a few years later. At Ignalina, the fuel was brand new, the reactor was stable, and the rods travelled down the entire height of the core, allowing boron to be introduced and the reaction to be brought back under control. This critical discovery was passed around the relevant nuclear Ministries and Institutes, but again nothing changed. Another KGB report dating from October 1984 highlighted complications with the cooling system experienced by Unit 1. The necessary information had been sent to the relevant Ministries at the time, *"but even on Units 5 and 6 that are now* [in 1984] *under construction, these comments are not taken into account."[92]* In light of all these repeated, wilful examples of negligence, I find myself agreeing in many ways with Chernobyl's Deputy Chief Engineer Anatoly Dyatlov, when he said years later that, *"the RBMK reactor was condemned to explode".[93]*

---

[92] Tykhyy, Volodymyr. *From Archives of VUChK-GPU-NKVD-KGB, Chernobyl Tragedy in Documents and Materials*. Security Service of Ukraine, 2001. Document 17: "Main engineering and technical faults of
Chernobyl NPP units, resulting from design".
[93] Dyatlov, Anatoly. *"Why INSAG Has Still Got It Wrong."* Nuclear Engineering International. April 8, 2006. Accessed March 02, 2016. http://www.neimagazine.com/features/featurewhy-insag-has-still-got-it-wrong.

CHAPTER THREE

Fascination

I can't remember when I first became interested in Chernobyl. As a child, I remember occasionally hearing snippets of stories about the city abandoned after a nuclear meltdown. I had no idea what a nuclear meltdown was, but to a child the phrase sounded like something out of science fiction. As fictional as it may have sounded, it wasn't the accident that piqued my interest, it was that an actual, real-life city was deserted somewhere. The idea blew my mind. I have often imagined what it would be like to walk through such a place, to be somewhere so familiar and yet so empty; to wonder what it was like before whatever tragedy befell it.

It wasn't until I attended university in 2005 and saw a collection of photographs taken by a biker riding alone through the Exclusion Zone (although her story turned out to be a fabrication), long before going there was popular, that I became fascinated by what happened. I sought out as many photographs of the accident as I could, and that was when the iconic silhouette of Chernobyl's ventilation stack became ingrained into my memory. In 2007, the dark, sprawling PC video game 'Stalker: Shadow of Chernobyl' was released, allowing me to visit and explore - in a manner of speaking - the places I had seen and read about. The game is set in an alternate timeline where strange, supernatural anomalies have appeared across the Exclusion Zone following the Chernobyl accident. While it has its shortcomings, the Ukrainian developers recreated many recognisable locations with photo-perfect accuracy and the setting was dripping with atmosphere. The more I played, the more I longed to go there and see the plant for real. Still, as a student, a lot was happening in my life and I soon moved on to other, equally fascinating things. Over the years, I returned to the story of what happened a few times, and each time I felt a greater desire to learn more.

Fukushima changed everything. On March 11th, 2011, at 14:46 JST, a magnitude 9.0 earthquake - the 5th strongest ever recorded - occurred 70 kilometers east of the Oshika Peninsula of Tōhoku, Japan. The undersea quake caused a tsunami as high as 40 meters to hit the coast, devastating everything in its path and travelling up to 10km inland. Over 16,000 men, women and children lost their lives in the ensuing chaos, and a further 400,000 lost their homes after over one million buildings were damaged or destroyed. [94] The World Bank's estimated economic cost was US$235 billion, making it the costliest natural disaster in world history.[95] The tsunami overwhelmed the inadequate flood defenses of the 40-year-old Fukushima Daiichi nuclear plant with ease and submerged the entire site, including its backup diesel generators. The instant the earthquake had been

---

[94] *Damage Situation and Police Countermeasures Associated with 2011 Tohoku District - Off The Pacific Ocean Earthquake*. Report. Tokyo: National Police Agency of Japan, 2016.
[95] *The Recent Earthquake and Tsunami in Japan: Implications for East Asia*. Report. World Bank, 2011.

registered offshore, Fukushima's three active reactors shut down and commenced decay heat cooling via their emergency diesel generators. Now those generators were underwater, useless, and the country was in turmoil. Fire trucks fought their way along roads upturned by the earthquake and tried to connect their hoses to the reactor pumps, only to find that no adaptors for the connection were available on-site. Despite the valiant efforts of Fukushima's staff, all three reactors melted down and their containment buildings were badly damaged by hydrogen explosions. It has become the world's second worst nuclear disaster, and the only accident to be rated a 7 on the 7-point International Nuclear Event Scale other than Chernobyl itself. Fukushima's other three reactors were offline for refuelling at the time of the accident, otherwise who knows what would have happened.[96]

After that fateful tsunami overran the Japanese plant, I sat glued to my computer, trawling the net for every piece of new information. Chilling phone-videos, uploaded to YouTube by survivors of the unstoppable, encroaching wall of water, were viewed with wide eyes, over and over again. It wiped out everything in its path. Vehicles, from simple bicycles to monolithic fishing vessels, were tossed inland like paper; whole towns were flattened and pushed inland. As the situation at Fukushima Daiichi worsened by the hour, residents of online forums and blogs speculated about what would happen. Would this be another Chernobyl? Armchair nuclear experts appeared from nowhere, offering opinions on nuclear safety systems and how Japan was well prepared for such an event.

As it turned out, the one person who appeared to me to be the most well-informed was wrong when he said the reactors

[96] Kitazawa, Koichi, Chairman. "The œ Fukushima Daiichi Nuclear Power Plant Disaster: Investigating the Myth and Reality. London: Routledge, 2014.

were borderline indestructible and that even this tsunami wouldn't cause a meltdown. Like many others, I wondered what the implications would be for the environment and the residents living nearby. I realised that - in spite of my interest - I didn't really, fundamentally understand how nuclear reactors worked, nor how good their safety systems were. The likes of Greenpeace are loud and uncompromising in their view that nuclear power is unsafe and produces damaging, non-disposable waste. Advocates reply that it causes proportionally far less deaths than coal, which accounts for 3 times the amount of electricity generated worldwide as nuclear; that fly ash emitted by a coal power plant carries 100 times more radiation into the surrounding environment than a nuclear power plant producing the same amount of energy; indeed, that nuclear generates more clean electricity than any other widely commercialised form of energy.[97]

So which is it? There's so much fear and propaganda surrounding nuclear power that it's almost impossible to know what to believe when you're uninformed. I wanted to learn the truth for myself, and that's when I became more serious about learning the secrets of nuclear power and its potential for harm. What better event to learn from than the worst man-made disaster in history? I wanted to know what had gone wrong at Chernobyl, how it had happened, who was responsible, how it was resolved, and what lessons were learned. First, I watched as many documentaries as I could find. Some appeared to be objective and informative, others were speculative - even brazen - with their invention of 'facts' about what happened. To confuse matters further, the original Soviet account of what

[97] Hvistendahl, Mara. "*Coal Ash Is More Radioactive than Nuclear Waste.*" Scientific American. December 13, 2007. Accessed March 03, 2016. http://www.scientificamerican.com/article/coal-ash-is-more-radioactive-than-nuclear-waste/.

happened was very misleading. This meant a lot of books written in the years after the accident were inaccurate. I came to realise that a lot of false information surrounds this legendary nuclear accident; everyone has heard of it, but few know what really happened. This blurring of information only made me more determined to learn the truth.

In late August 2011, I happened to be browsing a photography forum for the first time in months, when I saw a thread advertising a trip to visit the Exclusion Zone. It had been fully booked, but as crunch time neared there were a handful of drop-outs. Scheduled to depart on October 8th, it was mere weeks away. I knew that tour groups offered guides to show curious visitors around, though they were on hold because of vandalism, but that these tours followed an approved, supervised route. This wouldn't be like that: the group was expecting to have unrestricted access to Pripyat. I didn't know anyone going, but I decided then and there that – 26-years-old, penniless and unemployed – I had to join that expedition. At £425, plus transport to Ukraine and evening meals, the cost was less than I expected; an achievable target. Of course, the cost of first getting to London from where I lived in Aberdeenshire, Scotland, and then flying to Kiev and back doubled the price to somewhere close to £1000. The money would cover buses, accommodation, guides, breakfasts, and – most crucial, I suspect – bribes.

How was I going to come up with £1000 in just a few weeks? I decided to sell my first proper electric guitar, a beautiful transparent red Ibanez Joe Satriani Signature JS-100, and an excellent Nikon 105mm macro lens I used nowhere near enough to justify its £650 value. I was sad to see the guitar go. It was the first instrument I had ever loved, but I'd replaced it a year earlier by a 30th Anniversary Schecter C-1, and the lens I used perhaps once every few months. I put them both on eBay. Two African scammers and several wasted weeks later and I had the money I needed, thanks to a generous loan from my parents making up the shortfall.

The group was to assemble and set off on October 8th from Luton Airport outside London, then fly to Borispol Airport near Kiev in Ukraine, where we'd meet up with more people from around Europe. First I had to reach London from the old stone mill house where I lived in the countryside north of Aberdeen, about as far away from London as you can get in Britain. Faced with the choice of a hellish, twelve-hour bus marathon or a two-and-a-half-hour railway jaunt down to Edinburgh, followed by an overnight sleeper express to London, I chose the train. I'd wanted to travel on a sleeper train ever since I was a boy. It sounded so adventurous (Murder On The Orient Express, anyone?) and had the added benefit of getting a proper rest, which just wouldn't happen on the cramped, uncomfortable bus.

On Friday evening my father drives me to the nearest bus stop, five miles from home, and bids me farewell. One hour and 50 kilometers later I walk into Aberdeen's elegant Victorian station, with its recently refurbished wrought iron and glass ceiling, and board the first of my two trains. The journey down Scotland's east coast is uneventful, and I soon can't see anything in the windows except my own reflection, so I recline in my seat, pull out my phone and load up Minecraft: Pocket Edition. It came out today, and for some strange reason I'm excited at the prospect of becoming the first person to ever play Minecraft at Chernobyl. Having crossed the majestic Forth Rail Bridge in total darkness, the first leg of my journey ends at Edinburgh Waverley. It's 11pm. I disembark and find my next train idling in a quiet corner at the opposite end of the station, where I check with the uniformed conductor that it's going to London.

I once boarded a 9-car Virgin Pendalino for the short 25 kilometer hop from Preston to Lancaster, only to realise after half an hour that we hadn't stopped. Upon being asked, the unimpressed conductor struggled to maintain his poker face while informing me I was on a non-stop express to Glasgow - almost 300 kilometers away. Oh. They diverted the train to make a brief stop halfway at Carlisle, just for me.

Not today, she reassures me. *"We're full up tonight."* I find my cabin and open the door. The other chap isn't here yet so, childishly, I claim the top bunk as my own, planting my bag down like a flag. Time passes but nobody else arrives, and as we're about to leave, the same lady knocks on the door, pokes her head in and declares he must not be coming. I'll have the cramped compartment to myself, although I soon discover that sleeping on a train isn't so easy. It constantly rattles and rolls, stops and starts, as I speed south towards the capital.

It's 4am before I know it, and the train is easing into London. I'm cold and tired, but after a frigid walk between stations I'm soon on the next two-hour train to Gatwick Airport. Having travelled the farthest to get here, I'm also the first of the group to arrive, but by 9am others are starting to appear. I approach the assembled men and women and introduce myself. It's nice to finally put faces to the names of the people I've been talking to for the last couple of weeks. I meet a lot of great people today, but in particular I meet Danny, Katie and Dawid. The four of us will stick together for the rest of our journey.

A disinterested announcer informs us that our plane is ready for boarding, and we walk out across the tarmac to the waiting Ukraine International Airlines Airbus A320. I try to keep up a calm façade, but inside I'm panicking; I've only flown twice before - at night - and hated it. The possibility of being in a plane crash, powerless to prevent what's about to happen, has always terrified me and is a routine nightmare of mine. A window seat behind the port-side wing affords me an excellent view, but my phone is a better distraction from my nerves until the flight attendant orders all gadgets to be switched off. I close my eyes to block out my surroundings as I'm forced back into my seat by the jet's powerful engines. It's as thrilling yet terrifying as I remember.

The view from a plane is better than I ever imagined; seeing the world from this height for the first time makes me realise how truly insignificant we all are, as cliché as that is. I spend half the flight trying hard to guess where we are from visible coastlines, and equally hard trying not to think about the 35,000 feet between me and the ground. The aircraft begins its bumpy descent through darkening clouds into Borispol Airport in late afternoon, after four and a half hours in the air. It's overcast and raining, but I don't care - I'm back on solid ground and can forget my fear of flying for the time being. How do aircrews do it?

It's obvious our group stands out, because as soon as we enter the terminal people all around us are staring. We've been instructed ahead of time that under no circumstances are we to tell the airport staff at Borispol why we've come to Ukraine. Instead, we say we're tourists on a photography trip. The skinny, blank-faced man in the booth stares at me, skeptical. Do all foreigners come to Ukraine for Chernobyl? I doubt it, but I flash him a brief, innocent smile, just in case. Apparently, if they knew our true intentions there's a chance we wouldn't be allowed into the country, though I'm not sure why.

We have a few hours to kill. A bus will pick us up at 8pm, but until then we're free to pass the time. After exchanging some currency, I join Danny, Katie, Dawid and a friendly guy named

Josh in a search for food. Like the gormless tourists we are, we settle on the first familiar site - a small, American-style restaurant in the main terminal building, decked out like a classic 50s diner. The walls are papered with old black and white photographs of New York, complemented by hanging prints advertising Coca-Cola. The menu is styled after a front page of The Times newspaper. We're famished, but since none of us apart from Dawid speak or read the slightest bit of Ukrainian, and the waitress doesn't speak or read English, we each settle for tea. I guess tea is universal.

Sipping our piping hot green tea, my new friends and I chat about Chernobyl, our camera gear, where we're all from, and how excited we are to be here. The time flies by, and before long we're boarding our coach to the thousand-year-old central city of Bila Tserkva 80km away, where we'll spend the night before pushing on to an ICBM museum in the south. We arrive at Bila Tserkva without incident by 11pm, the only notable view on our darkened approach to the hotel is an intriguing floodlit industrial site. After standing around in a hotel lobby for 20 minutes while our guides have a lengthy discussion with hotel staff, we're directed up a marble-and-stained-glass staircase. I get the impression they didn't know we were coming. On the top floor we again find ourselves with no direction, until Dawid comes to the rescue and explains the situation through a mixture of hand gestures and Polish to a cleaning lady. Once everyone has, at last, dropped their belongings in their rooms and explored the building (the roof access, our natural first port of call, is locked), a collective decision is made to retire to the hotel pub.

Collective, that is, except for me. I'm as tired as everyone, but I didn't travel all this way to stand around and get drunk - I want to explore. Dawid agrees to accompany me after some badgering, so we each grab a tripod and camera and head out into the night. Our hotel lies on the northern edge of a well-lit intersection along with a couple of shops and restaurants, but beyond it the streetlights thin out, leaving long stretches of overgrown sidewalk and potholed road in darkness. Dawid and I say little as we retrace my memorised route to the industrial site we passed earlier. Along the way, I encounter my first unexpected sight - stray dogs. We've only been walking for ten minutes but already two or three have walked past, ignoring us on their blithe night wander. This may not be so unusual to some, but stray dogs just aren't something you see in the north of Scotland. To counter the dogs, it isn't much longer before I encounter my first *expected* sight - a Lada Riva; one of the most iconic vehicles of the Soviet Union.

The white central building of the industrial site looks like a grain silo, comprised of two sets of twelve featureless silos separated by a tall building in the middle and two huge silos at the far end, all connected by a flimsy-looking horizontal section. Dawid and I photograph it from the shadow of a tree, trying to keep out of sight of the man in an ex-army truck sitting out front. We don't hang around long, only moving a little further down the road to photograph the site's boiler plant before heading back to the hotel to sleep.

The Strategic Missile Forces Museum was once a top secret Soviet missile base, used to house the cold-launched SS-24 "Scalpel" silo-based missile. On display, among 2000 other items of interest, is the 35-meter long, much feared SS-18 "Satan" intercontinental ballistic missile. It had the highest yield of any nuclear missile ever developed, at 20 Megatons, and was far more powerful than any ICBM in service today. For comparison, the Hiroshima bomb was 'only' 16 Kilotons to the SS-18's 20,000 Kilotons, which has an area of destruction of 800 square miles. Following the collapse of the Soviet Union, all missile bases in Ukraine were demolished as part of the 'Strategic Arms Reduction Treaty' (START) agreement with the USA. All except this one, which was turned into a museum. The base is fun - I explore a 12-storey missile command module buried 40 meters underground, photograph lots of exotic military vehicles and see some impressive missile technology up close, but it rains the entire time and isn't what any of us travelled all this way to Ukraine to see. We're itching to visit Chernobyl.[98]

We depart from the museum at around 2:30pm and begin the ten-hour slog towards the town of Slavutych, which will serve as our base of operations over the next few days. As the light outside grows dim, I fight the tedium by taking light-trails photographs of passing vehicles through the window. Soon, equally bored, almost everyone on the bus joins in. We pass through Kiev, not really seeing anything apart from distorted, rain-soaked shapes and the enormous, floodlit, 102-meter tall Motherland statue, standing guard atop the city's tallest hill. Beyond Kiev's city boundaries, the dead-straight, well-worn road is pitch black. There are no streetlights and other vehicles are few and far between; all I see beyond the bus's own faint glow are ghostly silhouettes of a corridor of trees. Lacking anything better to do, I burn an hour or so explaining exactly what happened at Chernobyl to Danny, Katie and Dawid. At one point during our journey, the bus appears to spontaneously catch fire, alarming everyone except the driver. We smell burning and see smoke in the cabin, but he's unphased and keeps driving as if this is perfectly normal. I'm beginning to appreciate how nonchalant Ukrainians are.

---

[98] Strategic Missile Forces Museum. Accessed March 03, 2016. http://www.rvsp.net.ua/.

After ten lifeless, interminable hours, we arrive in Slavutych. Erected 50 kilometers east of Chernobyl, Slavutych began construction in 1986 shortly after the accident, specifically to house Chernobyl's workers and their families after Pripyat was rendered uninhabitable. Its name comes from the Old Slavic name for the nearby Dnieper River. The town is home to 25,000 inhabitants, and its economic and social situation is still heavily influenced by the power plant and other Chernobyl zone installations, because most of the residents either worked or still work there. Its construction involved architects from 8 different Soviet republics, and, as a result, the city is split into 8 distinct areas - each with their own different styles of architecture and colour schemes. Despite being very modern in comparison to other places in Ukraine, there has been a high rate of unemployment since the power station shut down its last reactor in December 2000, leaving only 3,000 residents employed there.

We're told to split ourselves into groups, so my friends and I elect to get a 4-bedroom place together. The bus creeps around Slavutych in the dark, dropping groups off here and there until it's our turn. We're deposited outside a 5-storey building, where a short, plump, dark-haired woman in her early 40s awaits. She gestures for us to follow, then leads the way upstairs to a five-room apartment on the top floor. It's her own home! Dawid, being Polish, understands bits and pieces of Ukrainian, and infers that she's renting it out to us as a way of making a little extra money, and is living with her children in her mother's flat across the hall for the duration of our stay. It's a lovely little place, very homely and warm, with family photographs lining the walls and soft toys in the bedrooms; far more comfortable and welcoming than any hotel. I feel guilty about the arrangement, but try to reassure myself that it's to the benefit of all involved. We settle ourselves in, make several cups of her delicious tea and chat for a while, but soon drift off to our beds in anticipation of the days ahead.

CHAPTER FOUR

The Accident

On April 26th, 1986, just after 1am, a test was about to commence at Chernobyl's Unit 4 reactor. What followed was the worst nuclear disaster in history. That night, the shift comprised of 176 men and women at the plant, along with 286 construction workers building Unit 5, a few hundred meters to the southeast. Unit 4's control room operators, along with a representative of Donenergo - the state-owned electricity supplier and designer of the plant's turbines - were testing a safety feature intended to allow the Unit to power itself for around a minute in the event of a total power failure.

The principal concern of a nuclear reactor - particularly an RBMK reactor, because of its graphite moderator - is that cooling water continuously flows into the core. Without it there could be an explosion or meltdown. Even if the reactor is shut down, the fuel within will still be generating decay heat, which would damage the core without further cooling. Pumps driving the flow of water rely on electricity generated by the plant's own turbines, but in the event of a blackout the electrical supply can be switched to the national grid. If that fails, diesel generators on site will automatically start up to power the water pumps, but these take about 50 seconds to gather enough energy to operate the massive pumps. There are six emergency tanks containing a combined 250 tons of pressurised water which can be injected into the core within 3.5 seconds, but an RBMK reactor needs around 37,000 tons of water per hour - 10 tons-per-second - so 250 tons does not cover the 50 second gap. [99] [100]

Thus: the test of a 'run-down unit'. If a power failure occurred, the fission reaction would still be producing heat, while the remaining water in the pipes would continue its momentum for a short time and therefore steam would still be produced. In turn, the turbines would still rotate and generate electricity, albeit at an exponentially falling capacity. This residual electricity could be used to drive the water pumps for a few vital moments, giving the diesel generators sufficient time to get up to speed and take over, and it's the hardware behind this that was being tested.

Despite initial Soviet claims that the experiment had been intended to test a brand new safety system, this run-down unit is actually a standard feature of the RBMK design, and should have been made operational during Unit 4's commissioning three years earlier. In order to open the plant ahead of schedule, Chernobyl's Plant Manager Viktor Bryukhanov, along with members of various Ministries involved with the construction and testing of a new plant, signed off on safety tests that were never conducted, with the unwritten promise of completing them later. As reckless as it sounds, this was fairly routine practise in the USSR, as completing work ahead of schedule

[99] Medvedev, Zhores A. The Legacy of Chernobyl. Oxford: Basil Blackwell, 1990. Page 15-16.
[100] Medvedev, Zhores A. The Legacy of Chernobyl. Oxford: Basil Blackwell, 1990. Page 236.

entitled everyone involved to significant bonuses and awards. The hardware required precise calibration and revisions, and the test had already been conducted three times before on Unit 3 - in 1982, 1984 and 1985; all failed to sustain sufficient voltage - but engineers had by now made additional alterations to the voltage regulators, and so it was to be attempted again. The run-down test was originally scheduled for the afternoon of the 25th, but Chief Engineer Nikolai Fomin was asked by Kiev's national grid controller to delay it until after the evening peak electricity consumption period had ended.[101] The afternoon staff had been briefed on the test and knew exactly what to do, but their shift ended and they went home. Evening staff took over, but then they too left, leaving the relatively inexperienced night crew - who had never conducted a test before - the responsibility of starting a test they were not prepared for and had not anticipated doing.

To make matters worse, Unit 4 was at the end of a fuel cycle. One of the features of the RBMK design is 'online refuelling', which is the ability to swap out spent fuel while the reactor is at power. Because fuel burn-up is not even throughout the core, it was not uncommon for the reactor to contain both new and old fuel, which was usually replaced every two years. On April 26th, around 75% of the fuel was nearing the end of its cycle.[102] This old fuel had, by now, been given time to accumulate hot and highly radioactive fission products, meaning any interruption in the flow of cooling water could quickly damage the older fuel channels and generate heat faster than the reactor was designed to cope with. Unit 4 was scheduled for a lengthy shutdown and annual maintenance period upon

[101] Dyatlov, Anatoly Stepanovich. *Chernobyl: How It Was*. 2005.
[102] Medvedev, Grigoriĭ, and Andreĭ Sakharov. The Truth About Chernobyl. New York: BasicBooks, 1991. Page 34.

conclusion of the test, during which all of the old fuel would be replaced. It would have been far more sensible to conduct the test with fresh fuel, but management decided to push ahead anyway.

The test would involve inserting all 211 control rods part-way, creating a power level low enough to resemble a blackout while continuing to cool the reactor to compensate for fission products. Use the residual steam in the system to drive a turbine, then isolate it and allow it to run down, generating electricity through its own inertia. The electrical output would be measured, allowing engineers to determine whether it was sufficient to power the water pumps in an emergency. Because the deliberately low power levels would appear to be a power failure to the control computer, which would then automatically activate the safety systems, these systems, including the backup diesel generators and Emergency Core Cooling System (ECCS), were disconnected in order to re-attempt the test straight away if it proved unsuccessful. Otherwise, the ECCS would automatically shutdown the reactor, preventing a repeat of the test for another year. Astonishingly, these measures were not in violation of safety procedures when approved by a Deputy Chief Engineer, despite many subsequent reports to the contrary.[103] It's debatable how much of an impact these systems would have had on the outcome, but it was nevertheless a very foolish decision. Viktor Bryukhanov, along with Nikolai Fomin, who approved the test, paid the price with a sentence of 10 years' imprisonment in a labour camp and expulsion from the Communist party.[104] Countless others paid for it with their health and their lives.

---

[103] International Safety Advisory Group. *The Chernobyl Accident: Updating of INSAG-1: INSAG-7*. Vienna: International Atomic Energy Agency, 1992. Page 10.
[104] Karpan, N. V. *Trial at Chernobyl Disaster*. Report. Kiev, 2001.

There were problems from the outset. The test programme left for the night shift was full of annotations and hand-written alterations. A transcript of a telephone conversation between an unidentified operator and a colleague elsewhere in the building makes for scary reading: *"One operator rings another and asks, 'What shall I do? In the programme there are instructions of what to do, and then a lot of things are crossed out.' His interlocutor thought for a while and then replied, 'Follow the crossed out instructions.'"* [105] Then at 00:28, while reducing power to levels low enough to begin - a process which would take about an hour - Senior Reactor-Control Engineer Leonid Toptunov made a mistake when switching from manual to automatic control, causing the control rods to descend far more than intended.[106] Toptunov had only been in his current position for a few months, during which the reactor power had never been reduced.[107] Perhaps his nerves got the better of him. Power levels - supposed to be held at 1,500-Megawatts thermal (MWt) for the test - dropped all the way to 30MWt. (The reactor's output is measured in terms of thermal power - the turbogenerator's in electrical power. Energy is lost during the transfer from steam to electricity, hence the higher thermal figures.) Note that it was stated in the Chernobyl trial that power output dropped to zero, and specifically mentioned that the 30MWt figure was erroneous, but everything else I have ever read has said 30.[108] Either way, even 30MWt is near-as-makes-no-difference a complete shutdown and not even enough energy to power the water pumps. At such a low power, an atomic process of 'poisoning' the reactor begins - a release of the

[105] Legasov, Valerii. *"Moi Dolg Rasskazat' Ob Etom."* Pravda (Moscow), May 20, 1988. Quoted by Medvedev, Zhores A. *The Legacy of Chernobyl.* Oxford: Basil Blackwell, 1990. Pages 24-25.
[106] International Safety Advisory Group. *The Chernobyl Accident: Updating of INSAG-1: INSAG-7.* Vienna: International Atomic Energy Agency, 1992. Page 53.
[107] Medvedev, Zhores A. *The Legacy of Chernobyl.* Oxford: Basil Blackwell, 1990. p37. Quoting Shcherbak, Yuri. *Chernobyl,* Page 223.
[108] Karpan, N. V. *Trial at Chernobyl Disaster.* Report. Kiev, 2001. Page 37.

isotope xenon[135], which absorbs and seriously inhibits the fission reaction - and the test was over before it began. Had this massive drop in power never happened, the test would have proceeded without incident and the RBMK's dangerous shortcomings may never have come to light. Crucially, however, the man in charge of the test, 55-year-old Deputy-Chief Engineer Anatoly Dyatlov, did not stop.

Dyatlov was born into a poor family in central Russia. Through tireless hard work and a determination to do more with his life than his parents, he grew into an intelligent, self-made young man, and in 1959 graduated with Honours from Moscow's National Research Nuclear University. His work background, before moving to Chernobyl in 1973, involved installing small VVER reactors into submarines near Russia's eastern shores.[109] He was also, however, a man privately disliked by his subordinates due to a short temper, a low tolerance for mistakes and a tendency to harbour resentments.[110] Dyatlov had been present earlier in the day when the test was postponed; his patience was running short.[111] Instead of accepting that it was futile to continue, he reportedly went mad and rushed around the control room shouting. He did not want another test wasted and his reputation tarnished - he ordered the operators to recover the reactor and bring it back up to power. Continuing the experiment after falling to such a low power level resulted in the reactor becoming unstable enough to explode, and Dyatlov holds all the blame for this one crucial decision.[112] His behaviour could be, in part, because no nuclear plant operators in the Soviet Union knew of previous accidents at other nuclear

---

[109] Dyatlov, Anatoly Stepanovich. *Chernobyl: How It Was*. 2005.
[110] Medvedev, Grigoriĭ. *Chernobyl Notebook*. Moscow: Novy Mir, 1989. Chapter 2.
[111] Read, Piers Paul. *Ablaze: The Story of Chernobyl*. London: Secker & Warburg, 1993. Pages 78-79.
[112] Dyatlov maintained afterwards that he wasn't present when the power dropped, nor when the decision to continue was made. This contradicts several other eyewitness testimonies.

facilities, of which there were many. Authorities covered up all fatalities, while claiming in public that the technology was infallible - the best in the world. At worst, it was believed that an RBMK could only suffer a rupture of one or two water lines; an explosion was laughable.

Toptunov considered Dyatlov's decision to continue after such a massive drop in power to be a violation of safety procedures, so refused to comply, as did Unit Shift Chief Alexander Akimov.[113] Akimov was Russian, like most senior staff at the plant. Born on May 6th 1953, in the country's third-largest city, Novosibirsk, he graduated from the Moscow Power Engineering Institute in 1976 with a degree in thermal power automation processes, before moving to the Chernobyl plant in 1979 as a turbine engineer.[114]

Dyatlov grew angry and informed them that if they were not willing to do it, he would find someone who was. Akimov and the relatively inexperienced Toptunov, only 26 years-old, relented, and the test continued. Remember that a nuclear plant operator was a prestigious career with its own perks, and the possibility of losing that would have been a serious threat. Not only that, but Dyatlov may well have been the most experienced nuclear engineer at the entire plant. Even Chief Engineer Fomin was an electrical engineer - a specialist in turbines, like Bryukhanov. They respected his knowledge.

By 01:00, after around half an hour, the pair had succeeded in increasing the power to 200MWt by retracting about half of the control rods, but that was as high as it would go - nowhere near the intended 700MWt. Xenon poisoning had already taken its toll, seriously reducing the fuel's reactivity. Russian safety

[113] Medvedev, Grigoriï. *Chernobyl Notebook*. Moscow: Novy Mir, 1989. Chapter 2.
[114] *"Heroes - Liquidators."* Чорнобильська АЕС. Accessed March 20, 2016. http://chnpp.gov.ua/en/component/content/article?id=82.

regulations have since changed to require that an RBMK reactor be kept at a minimum of 700MWt during normal operation because of thermal-hydraulic instability at reduced power. Knowing 200MWt was still far too low to perform the test, they overrode additional automatic systems and manually raised still more control rods to compensate for the poisoning effect.[115] At the same time, they connected all 8 main circulating pumps and increased the flow of coolant into the core, up to around 60,000 tons per hour.[116] This volume of water was another violation of safety regulations, since very high water flow could lead to cavitation in the pipes. Increased coolant levels meant less steam, which soon caused the turbine speeds to drop. To counteract negative reactivity from all the extra coolant water, the operators withdrew most of the few control rods still inside the reactor, until the equivalent of only 8 fully inserted rods remained.[117] The normal absolute minimum allowed at the time was 15, which increased to 30 after the accident.[118]

The automatic safety systems would have, under normal circumstances, shut the reactor down a few times by now. At 01:22:30, minutes from disaster, Toptunov noticed the computer readings demanding that the reactor be shut down.[119] He and his fellow operators were calm but concerned about the state of the reactor. *"At the control board before program execution, some nervousness was seen,"* reported Razim Davletbaev, Deputy Chief of the turbine hall, in the 1987 accident trial. *"Dyatlov repeatedly told Akimov: 'Do not procrastinate.'"*[120] I struggle to understand why

---

[115] Medvedev, Zhores A. *The Legacy of Chernobyl*. Oxford: Basil Blackwell, 1990. p29.
[116] International Safety Advisory Group. *The Chernobyl Accident: Updating of INSAG-1: INSAG-7*. Vienna: International Atomic Energy Agency, 1992. Page 18.
[117] Ibid. Page 54.
[118] G. Medvedev states in *Chernobyl Notebook* that it was 30 a couple of times. Given his knowledge, I find it difficult to believe that he didn't know it was 15.
[119] *The Accident at the Chernobyl Nuclear Power Plant and Its Consequences*. Report. Vienna: USSR State Committee on the Utilisation of Atomic Energy, 1986. Page 17.
[120] Karpan, N. V. *Trial at Chernobyl Disaster*. Report. Kiev, 2001. Page 28.

Dyatlov wanted to continue from here. The reactor was clearly unstable and not even close to the power levels required by the experiment, so they would not have been able to gather useful readings regardless of what happened. Had Dyatlov accepted the futility of continuing, his men could have shut the reactor down. He didn't: the test began.

I don't know for certain Dyatlov's reasoning behind this decision, but he *was* pressured from above to get it done. The experiment had failed so many times by this point that Bryukhanov and members of the Soviet Academy of Sciences were eager to see the matter concluded. It may be that Dyatlov did not care if the results were useful. He simply wanted to report that the test had been carried out. This is speculation, of course, but it would explain what appears to be irrational behaviour from an otherwise impeccably rational man.

At 01:23:04, turbine 8 was disconnected and began to coast down.[121] The operators still had no idea what was about to happen and began a calm discussion, remarking that the reactor's task was complete and they could start to shut it down.[122] Within seconds of cessation of steam flow into the turbine, the main circulating pumps began to cavitate and fill with steam, reducing the flow of valuable cooling water and allowing steam voids (pockets of steam where there should be water) to form in the core. A positive void coefficient was occurring: the absence of cooling water causing an exponential power increase. In simple terms, more steam = less water = more power = more heat = more steam. Because 4 of the 8 water pumps were running off the decelerating turbine, less and less water was supplied to the reactor as power increased.

---

[121] International Safety Advisory Group. The Chernobyl Accident: Updating of INSAG-1: INSAG-7. Vienna: International Atomic Energy Agency, 1992. Page 54.
[122] Karpan, N. V. *Trial at Chernobyl Disaster*. Report. Kiev, 2001. Page 24.

Toptunov spotted this and shouted a warning to Akimov, who had direct control. He saw no choice: job or no job, the reactor was in a dangerous, unstable state and had to be made safe.[123]

At 01:23:40 on April 26th, 1986, 32-year-old Alexander Akimov made his fateful decision and announced that he was pressing the EPS-5 emergency safety button to initiate a SCRAM, causing all remaining control rods to begin their slow descent into the core.[124] [125] It was a decision that would change the course of history. (Note that Dyatlov claimed afterwards the EPS-5 button was pressed in a calm environment, purely to bring the test to a successful conclusion and with no concern about reactor readings, but I find that very difficult to believe under the circumstances and it goes against the testimony of other witnesses.) An emergency shutdown was Akimov's obvious choice. A large part of the reason why the core was so unstable was that almost all 211 rods had been removed, after all, leaving him and his colleagues with very little control over the reactor. He may even have considered this to be his only choice, given how many safety systems had been disabled, and with power increasing it's quite possible that there would have been a severe accident either way. Alas, it was, in fact, the worst thing he could have done. Within seconds, the control rods stopped moving.

Throughout the building, 'knocks' were heard from the direction of the main reactor hall. Akimov's control board indicated that the rods hadn't moved far before freezing, only 2.5 meters from their raised position. Thinking quickly, he released the clutch on their servomotors to allow the heavy rods

---

[123] Medvedev, Grigoriï. *Chernobyl Notebook*. Moscow: Novy Mir, 1989. Chapter 3.
[124] There seems to be some indecision between the various first and second-hand sources about exactly who pushed the button, Toptunov or Akimov. Far more state Akimov than Toptunov, and this does seem more likely.
[125] International Safety Advisory Group. *The Chernobyl Accident: Updating of INSAG-1: INSAG-7*. Vienna: International Atomic Energy Agency, 1992. Page 55.

to fall into the core by their own weight. They didn't move: jammed. *"I thought my eyes were coming out of my sockets. There was no way to explain it,"* recalled Dyatlov, six years later. *"It was clear that this was not a normal accident, but something much more terrible. It was a catastrophe."*[126]

Akimov didn't understand what was happening either. He, like the other poor operators in the control room, was unaware of a devastating fatal flaw in the reactor's design. While around 5 meters of each control rod was composed of the neutron-absorbing element boron to halt the reaction, the ends of each rod were made of graphite - the same reaction-*increasing* moderator as was used throughout the core of the RBMK. Between the graphite and boron was a long hollow section. The purpose of the graphite tips was to displace cooling water (which is also a moderator, albeit weaker than graphite) in the rod's path, thus increasing the boron's dampening effect on the fuel.[127] The moment all those graphite tips began to move inside the reactor, there was a surge in positive reactivity in the lower half of the core, resulting in a huge increase in heat and steam production. This heat fractured part of the fuel assembly, distorting the rod's tubes and causing a jam. When a control rod is fully inserted, the tip extends below the core, but now over 200 were lodged in the centre.

While the RBMK's designers were unaware of this flaw when the RBMK was first created, they had, they later admitted, forgotten to mention it, *"out of absentmindedness,"* once they realised.[128] I do not understand at all how such an obvious

---

[126] Dobbs, Michael. *"Chernobyl's 'Shameless Lies"* Washington Post. April 27, 1992. Accessed March 04, 2016. https://www.washingtonpost.com/archive/politics/1992/04/27/chernobyls-shameless-lies/96230408-084a-48dd-9236-e3e61cbe41da/.

[127] *The Accident at the Chernobyl Nuclear Power Plant and Its Consequences*. Report. Vienna: USSR State Committee on the Utilisation of Atomic Energy, 1986.

[128] Dyatlov, Anatoly. *"Why INSAG Has Still Got It Wrong."* Nuclear Engineering International. April 8, 2006. Accessed March 02, 2016. http://www.neimagazine.com/features/featurewhy-insag-has-still-got-it-wrong. I have seen the original source for this, but can't find it again.

design flaw can be overlooked by so many people. It blows my mind that the very system intended to prevent a fission reaction *increases* it in the most severe of emergencies - instances necessitating that the Emergency Protection System button be pressed - because the first stage of the designed emergency response is to introduce a moderator to the core. Anyone who knew anything about fission should have foreseen that this was clearly not how control rods should be designed. It is so obvious, in fact, that I am forced to conclude that I have overlooked a critical piece of engineering information, because no intelligent, rational person would have created such a system.

Within 4 seconds, the reactor's energy output had soared to several times its intended capacity. Runaway heat and pressure deep inside the core ruptured fuel channels, then water pipes, causing the pumps' automatic safety valves to close. This stopped the flow of coolant, increasing the rate at which steam was forming from the core's diminishing water supply. The reactor's own safety valves attempted to vent the steam, but the pressure was too great and they, too, ruptured.

At that moment, remarkably, there was a man in the expansive reactor hall of Unit 4 who witnessed all this.[129] Night Shift Chief of the Reactor Shop Valeriy Perevozchenko saw the top of the reactor - a 15-meter-wide disk comprised of 2000 individual metal covers which cap safety valves - begin to jump up and down. He ran. The reactor's uranium fuel was increasing power exponentially, reaching some 3,000°C, while pressure rose at a rate of 15 atmospheres per second. At precisely 01:23:58, a mere 18 seconds after Akimov pressed the SCRAM button, steam pressure overwhelmed Chernobyl's incapacitated fourth reactor. A steam explosion blew the 450-ton upper biological

---

[129] Medvedev, Grigoriï. *Chernobyl Notebook*. Moscow: Novy Mir, 1989. Chapter 3.

shield clear off the reactor before it crashed back down, coming to rest at a steep angle in the raging maw it left behind. The core was exposed.[130]

A split second later, steam and inrushing air reacted with the fuel's ruined zirconium cladding to create a volatile mixture of hydrogen and oxygen, which triggered a second, far more powerful explosion.[131] Fifty tons of vaporised nuclear fuel were thrown into the atmosphere, destined to be carried away in a poisonous cloud that would spread across most of Europe. The mighty explosion ejected a further 700 tons of radioactive material - mostly graphite - from the periphery of the core, scattering it across an area of a few square kilometers. This included the roofs of the turbine hall, Unit 3, and the ventilation stack it shared with Unit 4, all of which erupted into flames. The reactor fuel's extreme temperature, combined with air rushing into the gaping hole, ignited the core's remaining graphite and generated an inferno that burned for weeks. Most lights, windows and electrical systems throughout the severely damaged Unit 4 were blown out, leaving only a smattering of emergency lighting to provide illumination.[132] [133] [134]

"There was a heavy thud," remembered engineer Sasha Yuvchenko in a 2004 interview with the Guardian newspaper. He was only 24 years-old in 1986. *"A couple of seconds later, I felt a wave come through the room. The thick concrete walls were bent like rubber. I thought war had broken out. We started to look for Khodemchuk, but he had been by the pumps and had been vaporised. Steam wrapped around everything; it was dark and there was a horrible hissing noise. There was no*

---

[130] This and the following paragraph describing the explosion combine information from a variety of sources.
[131] There is some evidence to suggest the second explosion was nuclear in nature, but it is generally accepted to have been hydrogen.
[132] Medvedev, Grigoriĭ. *Chernobyl Notebook*. Moscow: Novy Mir, 1989. Chapter 3.
[133] Medvedev, Zhores A. *The Legacy of Chernobyl*. Oxford: Basil Blackwell, 1990. Pages 31-32.
[134] International Safety Advisory Group. *The Chernobyl Accident: Updating of INSAG-1: INSAG-7*. Vienna: International Atomic Energy Agency, 1992.

ceiling, only sky; a sky full of stars." Yuvchenko ran outside to see what had happened. *"Half the building had gone,"* he says. *"There was nothing we could do."*[135] One man was killed in an instant: 35-year-old pump operator Valeriy Khodemchuk was unfortunate enough to be in the main circulating pump room when it was annihilated by the explosion. His body was never recovered, leaving him entombed inside Unit 4.

Measuring radiation is a convoluted exercise. Units have included the curie, becquerel, rad, rem, roentgen, gray, sievert and coulomb. The main unit used for measuring the exposure of ionizing radiation at Chernobyl in 1986 was a Roentgen. It's now outdated, but I'll use it for the remainder of this book to keep things simple and because almost all reported measurements from the accident were in roentgens. We are all constantly being exposed to radiation from a variety of sources, such as aircraft, rocks, some foods and the Sun, and a typical human is exposed to ordinary background radiation at a harmless dose of 23 microroentgens-per-hour (μR/h), or 0.000023 roentgens-per-hour (R/h). A chest x-ray will give you a dose of 0.8 roentgens; the annual dose limit for a radiation worker, set by the U.S. Nuclear Regulatory Commission (NRC), equates to 0.0028R/h; the NRC's limit for the public is 0.1 roentgens for an entire year; aircrews, who receive a higher dose than radiation workers because they work in the upper atmosphere, where protection from solar radiation is lessened, receive 0.3 roentgens/year.[136] [137] The radiation in Chernobyl's Unit 4 reactor hall was now at an instantly-lethal 30,000 roentgens-per-hour. 500 roentgens, received over the course of 5 hours, is a fatal dose. 400 is fatal in 50% of victims. Anything even approaching that will hospitalise

[135] Parry, Vivienne. *"How I Survived Chernobyl."* The Guardian. August 24, 2004.
http://www.theguardian.com/world/2004/aug/24/russia.health.
[136] http://www.radiologyinfo.org/
[137] Nuclear Regulatory Commission

you for months if you're lucky, or cripple you if you aren't. The volume and intensity of radioactive particles thrown into the atmosphere on that night was equal to 10 Hiroshima bombs, not including the hundreds of tons of reactor fuel and graphite that landed all over the plant.

Back in the control room, Akimov attempted to phone the fire brigade - who had responded immediately to the devastation and were already on their way - but the line was dead.[138] The explosion tore away water pipes used to supply coolant into the bottom of the core, preventing the reactor from being fed with water by the mangled pumps. Unfortunately, the operators did not realise this - or were in denial, given the horrifying consequences a reactor explosion would entail - and their lack of understanding lead them down the wrong course of action which served only to exacerbate the situation and throw lives away. Instead, Deputy-Chief Engineer Dyatlov became convinced the explosions had been caused by hydrogen in the Safety Control System's emergency water tank, and that the reactor must still be intact. Even though he had no real basis for this explanation - and if he had looked out of a window he would have seen that he was wrong - he acted on this belief for hours after. There can be no other reason for an otherwise intelligent and rational human being defying the obvious. His version of events was told to everyone who asked, including Bryukhanov's report to the Government in Moscow, and was believed for almost an entire day. Curiously, despite admitting that he at first thought the explosion had been caused by hydrogen in a water tank, Dyatlov later said, *"I don't know how [Bryukhanov] reached that conclusion [that the reactor was not destroyed]. He did not ask me if the reactor was destroyed - and I felt too nauseated to*

---

[138] Karpan, N. V. *Trial at Chernobyl Disaster*. Report. Kiev, 2001. Page 24.

say anything. There was nothing left of my insides by that time."[139] Did he lie/mis-remember? I don't know. This is a narrative contradiction I cannot explain.

Every man in the control room was shocked and confused; they believed they had done everything right, under the circumstances. Akimov, prevailed upon by Dyatlov that the reactor could be saved, tried to start the diesel generators then sent two young trainees - Viktor Proskuryakov and Aleksandr Kudyavtsev - to the reactor hall with instructions to lower the control rods by hand. He sent them to their deaths. Dyatlov spent the rest of his life regretting the moment. *"When they ran out into the corridor, I realized it was a stupid thing to do. If the rods had not come down by electricity or gravity, there would be no way of getting them down manually. I rushed after them, but they had disappeared,"* he said a few years before his death.[140] The trainees made it to the massive reactor hall, having navigated their way past destroyed rooms and elevators, and only remained in the vicinity for a minute - stunned by what they saw - but that was enough. They died a few weeks later. Returning to the Unit 4 control room, tanned deep brown by the massive dose of radiation they had absorbed, the pair reported that the reactor was simply no longer there. Dyatlov refused to believe them, insisting they were mistaken: the reactor was intact, the explosion had come from an oxygen/hydrogen mix in an emergency tank. Water had to be supplied to the core!

The men on duty - particularly Dyatlov - were exhibiting strong signs of a psychological phenomenon often associated with man-made disasters, known as groupthink. Described as

---

[139] Dobbs, Michael. *"Chernobyl's 'Shameless Lies"* Washington Post. April 27, 1992. Accessed March 04, 2016. https://www.washingtonpost.com/archive/politics/1992/04/27/chernobyls-shameless-lies/96230408-084a-48dd-9236-e3e61cbe41da/.
[140] Dobbs, Michael. *"Chernobyl's 'Shameless Lies"* Washington Post. April 27, 1992. Accessed March 04, 2016. https://www.washingtonpost.com/archive/politics/1992/04/27/chernobyls-shameless-lies/96230408-084a-48dd-9236-e3e61cbe41da/.

'the desire for harmony or conformity in the group [which] results in an irrational or dysfunctional decision-making outcome,' Professor of psychology Doctor James T. Reason believes groupthink was a significant factor in the behaviour of the Unit 4 operators. *"Their actions were certainly consistent with an illusion of invulnerability,"* he says, in reference to choices made during the hour preceding the explosion, though still relevant here. *"It is likely that they rationalised away any worries (or warnings) they might have had about the hazards of their behaviour."*[141]

Valeriy Perevozchenko, the 38-year-old who witnessed the reactor valve-caps jumping up and down, was the first person of any authority to realise and accept what had really happened. He grabbed a radiometer rated for 1000 microroentgens - far higher than any normal reading. It went off the scale. Unbelievably, apart from one buried under rubble and another locked in a safe, there weren't any devices for measuring anything higher at the plant, as the explosion had burnt out the powerful sensors around the building.[142] Even standard safety equipment was locked up and inaccessible.[143] He took a guess, estimating 5 roentgens-per-hour. Not even close. Taking charge, Perevozchenko ordered two men to go and search for several missing people. Together, they managed to find and rescue an unconscious Vladimir Shashenok from under a fallen girder. Shashenok, a young Automatic Systems Adjuster who had been monitoring pressure gauges, received deep thermal and radiation burns over his entire body when the explosion destroyed the room he was in. His two brave rescuers received serious

---

[141] Reason, James T., Dr. "*The Chernobyl Errors.*" Bulletin on the British Psychological Society 40 (1987): Pages 201-206.
[142] I'm unsure if the radiometer buried under rubble and the other one locked in a safe are one and the same device. Some accounts say it was in a safe that became buried under rubble, others imply they're different. Either way, the point is that there was a chronic lack of detection equipment available.
[143] Medvedev, Grigoriï. *Chernobyl Notebook*. Moscow: Novy Mir, 1989. Chapter 3, 4.

radiation injuries, including a radiation burn on one man's back where Shashenok's hand rested as he was carried out. Both miraculously survived the accident, despite one of them receiving far more than an ordinary fatal dose. Vladimir Shashenok, a father of two who celebrated his 35th birthday only four days prior, succumbed to horrendous injuries in hospital four-and-a-half-hours later, having never regained consciousness. He was the second and final man to die on the first day. When his wife saw him, she was shocked. *"It was not my husband at all, it was a swollen blister."[144]*

Perevozchenko, meanwhile, went off in search of the already-deceased Khodemchuk, wading through debris, picking up pieces of fuel and graphite with his bare hands as he struggled to find his friend in the darkness. After an exhausting search turned up nothing but rubble and warped metal, he resigned himself to the fact that his colleague was lost and started back to Unit 4. By now Perevozchenko was suffering from the effects of strong radiation, continually vomiting and lapsing into unconsciousness as he staggered towards the control room. When he finally made it, he reported that the reactor was destroyed to Dyatlov, who rejected his assessment. The operators were already feeding water to the core.

Radioactive reactor fuel and graphite lay everywhere. Part of the roof had collapsed into Unit 4's section of the turbine hall, setting turbine 7 on fire and breaking an oil pipe, which spread the fire still further and set the hall's roof alight. Falling debris had broken the pressure valve on a feed pump, which was gushing out boiling, radioactive water.[145] Men and women rushed past chunks of uranium fuel as they battled to contain

---

[144] Lisova, Natasha. *"Widows Recall the Painful Days After Chernobyl."* The Moscow Times. April 26, 2006. Accessed March 06, 2016. http://www.themoscowtimes.com/news/article/widows-recall-the-painful-days-after-chernobyl/205323.html.
[145] Medvedev, Grigoriĭ. *Chernobyl Notebook.* Moscow: Novy Mir, 1989. Chapter 3.

the blaze, isolate electrical systems, and manually open oil-drain and cooling-water valves. Many of these brave souls later died, unaware they had been running among pieces of reactor fuel. For their part, Akimov and Toptunov stayed at the plant after the morning shift relieved them from duty at 6am, choosing to join the desperate effort to salvage the situation. The pair decided water flow to the reactor must be blocked by a closed valve somewhere, so they went together to the half-destroyed feedwater room, where they opened valves on the two feedwater lines. Next, they moved to another room, where they stood knee-deep in a highly radioactive mixture of fuel and water for hours, turning half-submerged valves by hand until the radiation drained their strength and they were evacuated to Pripyat's hospital.[146] Their noble efforts were in vain. The water lines had been destroyed along with the reactor - they were opening valves to nowhere - yet still the control room operators continued redirecting water towards the reactor even six hours after the explosion.

The Chernobyl plant staff were genuine heroes that night, in the true sense of the word. They did not flee when they could have. Instead, they selflessly stayed at their posts and replaced the hydrogen coolant in the generators with nitrogen, avoiding another explosion; they poured oil from the tanks of the damaged turbine into the emergency tanks outside, and spread water over the oil tanks to prevent more fire. Had none of this been done, fires would have spread down the entire 600-meter turbine hall and more of the roof would have likely collapsed. The flames would then have spread to Units 1, 2 and 3, which, in all probability, would have resulted in the destruction of all four reactors.

---

[146] Ibid. Chapter 4.

I would like to lift a paragraph straight out of Medvedev's Chernobyl Notebook, if I may, because it illustrates the gallantry on display that night. *"Aleksandr Lelechenko, protecting the young electricians from going unnecessarily into the zone of high radiation, himself went into the electrolysis space three times in order to turn off the flow of hydrogen to the emergency generators. When we take into account that the electrolysis space was alongside the pile of debris, and fragments of fuel and reactor graphite were everywhere, and the radioactivity was between 5,000 and 15,000 roentgens per hour, one can get an idea of how highly moral and heroic this [47]-year-old man was when he deliberately shielded young lives behind his own. And then, in radioactive water up to his knees, he studied the condition of the switchboxes, trying to supply voltage to the feedwater pumps. His total exposure dose was 2,500 rads [2,851 roentgens], enough to kill him five times. But after he had received first aid at the medical station in Pripyat, Lelechenko rushed back to the Unit and worked there several more hours."* This is just one example of one man's efforts. There are countless more I have omitted. What makes it so depressing is that a lot of what those men did to save the reactor only made the situation worse. They sacrificed their lives for nothing.

Even after he had gone back to work at the plant - and I cannot fathom how he found the strength - Lelechenko insisted he was fine and refused to go to hospital, instead returning home that evening to eat dinner with his wife. He hardly slept, yet still summoned enough energy to get up the following morning and go back to work, explaining to his wife, *"You can't imagine what's going on there. We have to save the station."*[147] He died two weeks later on May 7th, in a hospital in Kiev; the third victim of Chernobyl. He had been so sick that he would not have survived the flight to Moscow's specialist radiation hospital,

[147] Read, Piers Paul. *Ablaze: The Story of Chernobyl*. London: Secker & Warburg, 1993. Page 193.

where the others would soon find themselves. For his bravery, Lelechenko was posthumously awarded the Order of Lenin medal, the Soviet Union's highest national decoration.[148]

---

[148] *"Heroes - Liquidators."* Чорнобильська АЕС. Accessed March 20, 2016. http://chnpp.gov.ua/en/component/content/article?id=82.

CHAPTER FIVE

Arrival

I'm dragged from sleep by the incessant chirping of my alarm after a mere 5 hours, but am immediately eager to get moving. After years of waiting and countless hours scrutinising every facet of the accident, today is finally the day I'll see Chernobyl with my own two eyes. Bleary-eyed but alert, our foursome walks a few blocks to the restaurant where we've been told to regroup for breakfast. It's apparent that Slavutych is a colourful and more modern town than the other Ukrainian settlements we saw yesterday. The distinctive Soviet-bloc look to the architecture is still present, but it looks less dated somehow, as if more self-aware. Perhaps it's because the town went up as the Soviet Union itself fell down. The streets are wide and well maintained, with tall pine trees filling every available space between buildings and lining the roads. In a way it's a little too idyllic, a little too clean; I sense that the town isn't filled to capacity.

The restaurant lies on the northeast corner of Slavutych's main town square. One nondescript, white concrete building among many, it has no windows to peer through, so the four of us are initially unsure we have the right place. We step through the unmarked door, but even here there's just a short entranceway with a marble-tiled floor and staircase at the end - no people or furniture - so we ascend until we find a handful of recognisable faces. After depositing my equipment in a tall pile of bags by the stairs and grabbing a seat, I look around. This a rather surreal place to eat breakfast. The expansive hall - we occupy only a fraction - is draped in red and white, with chair covers and other décor more akin to a wedding than a light breakfast for a group of tired, foreign travellers visiting a nuclear disaster area. This must be where large town events are held. Behind the bar, four twenty-something girls in monochromatic blouses and skirts serve tea and coffee, under the watchful eye of a short, ample woman in her early 50s. Her thin smile and commanding presence remind me of a mafia boss.

I eat my fill of chicken, tomato and cucumber (summarising every single meal we'll be given in Ukraine) and guzzle as much tea as time and bladder permit, before we gather our belongings and set off for the train station. It's a dreary, wet morning, devoid of warmth, but we're lucky enough to encounter a break in the clouds and cover the distance to Slavutych station in 10 minutes. I observe the city's residents materialising from all around as they march in silence down the main road along with us; everyone heading in the same direction.

The train line gives life to Slavutych, like an artery from Chernobyl's heart. Without it, few of the 3,000 workers who continue to maintain the plant and study the Zone could do their jobs. There are no direct roads for cars or buses between them, and they certainly can't fly, so the train is their only realistic option. If Chernobyl was abandoned entirely, I think it's safe to assume that most – if not all – people living in this remote town would leave. Knowing this makes me melancholy, made more so because the town has a high level of radiation-related illnesses among its population. Until I arrived here, I had no idea the number of people still employed at the plant was so high. That so many people's livelihoods continue to rely on it after such a disaster, and in such conditions, gives you a renewed perspective and appreciation of your own circumstances.

Passing through a small but bustling market at the foot of the station building, we climb the cracked concrete steps to the nearest of the station's four open-air platforms. Nobody waits on the other three. Beyond the farthest platform are several long, white, two-storey (office?) buildings that look a little like converted sheds with their corrugated metal roofs. I don't see a soul through the unlit windows. With the train only minutes away, the platform is filling up. Unlike many of my fellow travellers, I pull out my camera and begin photographing the scene, eager to capture what I'm seeing, but quickly stop when I notice the eyes of the town's residents through the lens. They are not happy about being photographed. In fact, they're not happy that we're here at all.

An enchanting, old, grey Soviet electric train engine with cyan and magenta trim rattles into the station, pulling half a dozen carriages behind it. Without thinking, my years of mad dashes for seats on the daily rail commute from a previous job kicks in, and I slip through to the nearest door to hunt for a place to sit. Men and women already aboard don't exactly strain to hide their hostility towards me sitting near them, so I pick a pair of unoccupied inward-facing benches. It's only once we've squeezed four of us onto each padded bench that I realise we as a group, spread throughout the train, must be taking up an entire coach-worth of seats, leaving many legitimate passengers to stand. Despite not speaking the language, I can tell by the tones of their voices that they're justifiably angry. I suspect they see me as someone who has left his cushy life in his modern house to come and gawk at a reality they have to endure every single day, and I'm forced to admit that - relatively speaking - they're right. Even though I have a genuine interest - passion, even - in what happened here, probably more so than everyone I have arrived with, and perhaps even more than some of those who work here, I cannot deny that I'm better off than these people and could leave whenever I want. I know the sorry tales of Pripyat evacuees shunned by society out of an uneducated fear of radiation, many of whom were forced to return to the Zone, and an intense mixture of guilt and shame at my lack of courtesy wells up inside me. I won't sit in their seats again.

The train's deafening approach to the Chernobyl plant is breath-taking. During the first half you pass several farms and individual houses interspersed between woodland, cross over the Dnieper and Pripyat Rivers, and even pause for a few minutes on the outskirts of a village. The second half is flat marshland all the way to the horizon, and is an incredible sight. Though it feels somewhat perverse to admit, this is exactly how I imagined the landscape surrounding a nuclear disaster to look, and catch myself wishing it was misty. Of course, this is how northern Ukraine and Belarus have been since long before coal power was dreamt of, let alone nuclear, but at the very least it's fitting. It's autumn, so you would expect everything to be drab and retreating for the winter, but still I'm surprised at the dearth of variety in the landscape's colour and form. Beyond the occasional patch of pale green on a bush here and there, there's little visible life. We hurtle across the border and rattle through

15km of neighbouring Belarus, though there are no fences or markers to indicate the transition.

Piercing the horizon like a monument to caution, I catch my first glimpse of Chernobyl's 75-meter cooling stack as we turn a gentle corner a few miles out. It's gone again within moments as the track straightens and we bear down on the plant. Tension among my fellow travellers is building. Our train pulls into the station, inches forwards until its carriage doors synchronise with those of the enclosed platform, then comes to a halt. Doors open and its regular passengers disembark before we have a chance to move. I follow them out, but can only watch as they filter in silence through the only exit in the platform's far end. Where do we go? Nobody has told us anything. I can't see outside, the enclosed space is made from cold grey sheets of corrugated metal, split in two by a row of thick cyan pillars supporting a slanted roof. Looks like it was intended to be temporary. Our guide appears from the flock of workers evaporating into the complex and calls us over. We're ushered to an intersection between two corridors, where three stern, imposing, buzz-cut men in Army fatigues await. Two stand watch while a third guards a desk, armed with a clipboard. He takes his time calling names from a list and inspecting our passports, as each of us silently prays we haven't come all this way for nothing. Ten anxious minutes later, we're all approved without incident and are shepherded to the end of a low, wide corridor, where tube lights cast a golden glow onto more corrugated walls.

Our guide is Dr. Marek Rabiński. Sporting an untamed silver moustache, bald head and wide-rimmed glasses, he has the stereotypical look of an absent-minded genius scientist, and I like him immediately. He's the head of the Department of Plasma Physics and Technology at the Andrzej Soltan Institute for Nuclear Studies in Poland, a founding member of the Polish Nuclear Society, and an expert on the accident. Marek gives us a prolonged health and safety monologue before reciting our itinerary for the day, as if we hadn't already memorised it. Nobody is going to climb to the roof of a building and hurl themselves off, if we wanted to do that we could have done it back at home, but of course he's still required to warn us not to. Everyone, including me, is growing visibly impatient - muttering, tapping their feet, changing stance every few seconds and looking around. Now we're so close, it's almost painful to do nothing, like having your favourite food dangled just out of reach. We have precious little time here as it is. The preamble is slowed further by having to go through an interpreter - Marek doesn't speak English - until finally, after what feels like half an hour of itching to go, we step outside.

No longer a distant silhouette, I can pick out details and colours on the Sarcophagus, a few hundred meters away. It's partially obscured by a huge, decayed rolling crane from where I'm standing, but I ignore the obstruction and take a picture anyway, struggling for a better angle in the entranceway. True to expectations, rain begins to fall from the colourless sky, so I stash my camera and join the others climbing aboard a wonderful old red and white 1970s bus. It's the exact same type that ferried out evacuees after the accident. A soldier of indeterminate rank will be joining us. A low ranking officer, maybe? I can't tell, he has no visible insignia under his coat. With a regulation shaved head, wearing aviators and perpetually chewing gum, he's about 5' 8" - a little shorter than me - and speaks with an accent as thick as tar. I love it, he almost sounds like he's speaking English phonetically, it's that thick. Sadly, he rarely says anything, much like the gruff, crinkled bus driver, and both look like they could think of a million better things to do than babysit us.

I'm too excited to care. With everyone safely aboard, we're driven for 5 minutes around to Unit 4. Standing in front of it, I can see the Sarcophagus in all its terrible glory. It is *immense*! I knew it was big, of course, but I'd failed to comprehend just how truly prodigious it really is. The chimney reaches a height of 150-meters, which is difficult to visualise for someone raised in a hundred-year-old, two-storey stone mill house in the countryside. Incidentally, two years later I built a full-scale model of Chernobyl in Minecraft using a few plans I found online, and confirmed again that it is, in fact, gigantic.

I'm flooded with emotions; I don't know why it means so much to me to be here, but it does. After watching so many documentaries and dramatisations of the accident, and reading so much about the men and women involved, I feel overwhelmed to actually be standing where it all happened. Perhaps this is similar to how some people feel when they visit Auschwitz or the beaches of Normandy.

Still, the structure looks a little different from the way I'm used to seeing it. The focus of my research into the accident until now has only extended as far as 1987 - when the sarcophagus was first built. 25 years-later, its roof and western wall are now held up by a 63-meter-tall support framework known as the Designed Stabilisation Steel Structure (DSSS), which was completed in 2007 as part of the Shelter Implementation Plan (SIP) - a long term project for making the site safe for the future. The weight of the original shelter's roof had been supported by two huge metal beams resting and imparting serious stress upon what was left of Unit 4's western wall, which had been severely damaged by the 1986 explosion. By the early-2000s it had been in serious danger of collapse, so now the bright-yellow-and-grey DSSS uses cantilevers to take 80% of the roof's 800 tons off the wall, thus preventing a collapse.[149]

On a patch of well-tended grass 150m from the Sarcophagus, there's a stone memorial depicting a pair of cupped hands, supporting the building and its chimney. Its plaque reads: *"To heroes, professionals to those who protected the world from nuclear disaster. In honour of the 20th anniversary of shelter object construction."* The rain is getting heavier, but I keep on taking pictures of the crumbling Sarcophagus until we're ushered into the nearby information centre, just this side of a concrete and razor wire-topped wall. Inside the single cramped room is a wonderful cross-section model of Unit 4, giving you an accurate scale recreation of the destruction within. The pump room where Khodemchuk died is completely buried. To the right of the model is a wall of glass, affording me the closest and most detailed view of Chernobyl of the entire trip, but we're inexplicably forbidden from taking photographs from this perfect vantage point. I don't understand why; it's frustrating beyond belief. A suited official gives us a short talk on what's happening with the New Safe Confinement project (NSC), and the progress made so far. She mentions that the famous ventilation chimney will have to be dismantled before the NSC can be rolled into place in a few years' time. This has since happened, in February 2014.

Back outside, we assemble in front of the plant for a photo taken by our trip's organiser. I have a funny picture of him from this moment, where everyone gave him their cameras and he has about 20 DSLRs hanging from his neck. A booming sound

---

[149] *"The Chernobyl Shelter Implementation Plan."* European Bank for Reconstruction and Development. Accessed March 06, 2016. http://www.ebrd.com/what-we-do/sectors/nuclear-safety/chernobyl-shelter-implementation.html.

emanates from nearby and reverberates across the landscape, like that of a cathedral bell, half submerged and struck by a sledgehammer in a steady rhythm. Beyond the razor topped-wall, a construction crew use pile-drivers to dig foundations for the New Safe Confinement's rolling track. I'll hear this sound everywhere throughout the next two days. To me, it's the sound of the Zone.

Our bus trundles towards Pripyat. As we approach, a lone, bored soldier manning his checkpoint raises a simple barrier by hand, granting us passage through the perimeter fence. We're deposited on a road near the city centre and told to be back in 90 minutes. Danny, Katie, Dawid and I are joined by a couple of others, and together we split off from the main group and head towards Pripyat's tallest tower, which conveniently stands watch over the north-west corner. My first impression is exactly how I expected. Everything is here - street lights, road signs, a child's bicycle lying by the roadside - but it all speaks of a forgotten life. The lights lack bulbs; road signs are rusty, their markings faded; the bike's wheels and handlebars are missing. In all my years of exploring abandoned places, there's only one other site that approached such a complete feeling of lost community life. The Bangour Village Hospital, which opened near Edinburgh in 1906 on a 960-acre estate, was one of the first village-plan psychiatric hospitals in Scotland. It's been abandoned for over a decade, but its landscaped grounds are maintained to this day and you can still see the church, shop, streetlights, bus stops, road markings and all the other little details that you'd never normally think about. Pripyat has details like that on a massive scale.

We arrive at the 16-storey residential tower - named Fujiyama after Mount Fuji in Japan (no idea why) - after a brisk 10-minute walk. One exhausting climb later, during which I lament carrying far too much heavy equipment and begin to reevaluate my packing philosophy, I emerge onto the desolate roof. The view is astounding - an abandoned, overgrown city laid out before me, like in a dream. White and grey brutalist concrete structures, most devoid of any perceptible flourishes, protrude above what is essentially an untamed forest, while the distant, hazy silhouette of Chernobyl is just visible on the horizon through the mist. Dark clouds linger in the air, soaking everything, but it somehow seems appropriate. There really is no other feeling on Earth like being in this empty, crumbling, almost ineffable city. Standing here in silence, but for the wind in my ears, it's as if everyone on Earth perished long ago and somehow I survived - I can feel it in my bones. I suddenly feel an intense loneliness, despite the presence of my new friends. Do they feel the same? I don't ask.

We're conscious of our strict 90-minute time limit, so don't linger for longer than necessary. On the highest floor of the building, above the residential levels and among bare concrete, water tanks and pipes, we find the mummified corpse of a dog. Did it come up here seeking shelter or its missing owners? Holes cover its body. Bullet wounds? It may not have escaped the patrolling extermination squads after the evacuation, but holes could mean anything after 25 years. Radiation sickness is bad enough in a human being, but at least a person would be informed about symptoms and remedies. For an animal, which has no clue what's happening to it or why, nor why the humans who cared for it have vanished, those final weeks of life must have been unbearable. I hope the poor creature escaped the worst effects of radiation and merely starved.

On the trek back to our bus, we make a short stop in one of Pripyat's many nurseries. Empty cots and children's toys fill room after silent room - each colourful, painted wall displaying smiling animals, cartoon landscapes, numbers and the alphabet.

Once everyone has returned to the rendezvous point (briefly thinking we've lost one straggler), we head back out of Pripyat and towards the research buildings scientists use to monitor radiation levels throughout the Zone. On the way, we pass the infamous Red Forest, which turned from green to red because of the extreme amount of radiation passing through it, and drive through the ancient town of Chernobyl, from which the plant inherited its name. I wish I could recall what the scientist we met told us about the work they're doing, but I can't remember any of it because we were back to our everlasting translation process; in my frustration I quickly lost interest. Moving on, we make a couple of brief stops at other places of note. First, the vibrant white, gold and neon-blue St Elijah Church, the only remaining active church in the Zone. Run today by an Orthodox priest who is one of the town's few permanent residents, it is famous for somehow remaining relatively clean from radiation, even right after the accident, or so the legend goes. Next, an old harbour off the Pripyat River, where rusted, listing, radioactive ships struggle to stay afloat.

We stop for a few minutes at a memorial to the fallen firefighters on our way back to the plant, with its life-size sculptures of 6 brave men attacking the blaze. A lone doctor stands at the rear. I'm not sure which is more tragic, that those poor souls on the roof either didn't recognise the magnitude of what they were facing, or they did, and knowingly sacrificed themselves. I wonder how many knew that the rubble they stood on was radioactive fuel and graphite, that the air they breathed was poisoned with lethal radionuclides, rendering them dead men walking within minutes. Regardless, they stayed at their posts and fought almost 40 distinct fires, despite everything, and their sacrifice prevented untold devastation. The plaque on the memorial reads, with all seriousness, *"To those who saved the world."*

Our last stop before lunch is on an open stretch of road, about one mile southeast of the plant. We're afforded a magnificent view of the crippled Unit 4 and its sarcophagus in the distance. Across the river to my right is a half built cooling tower, along with the partially complete Unit 5, which had been due to open a few months after the accident. It was never finished; the men and women put down their tools and abandoned the cranes where they stood.

We arrive for a late lunch at the dining hall used by Chernobyl's staff. After stepping in a red liquid at the door, used to neutralise radioactive dirt that may have stuck to our boots, we wash our hands and climb the stairs into the hall. Other than kitchen staff, the place is more or less deserted, so we queue up and enjoy our most substantial meal for the duration of our trip. After lunch, the bus takes us for a closer look at Unit 5, surrounded by the rusting tower cranes that were in use even at the moment of the explosion. What I would give to go inside... My photos of the building are terrible. I'm too focused on staring at Unit 5 to find a decent vantage point, squandering my brief time here. As I make my way back to the bus through a small wooded area, dotted with unidentifiable scraps of machinery, I come across a handful of cute and playful stray puppies, apparently adopted by the soldiers stationed nearby. Were they descendants of the dogs that had lived here before the accident? They must be, I can't imagine military personnel being allowed to keep pets on duty. As we depart, I catch a glimpse of an enormous, black rolling crane - the same kind I've seen used in many photos of Chernobyl's construction. I curse myself for not spotting it earlier as it disappears from view.

Our limited time is nearing its end as we stop beside the plant's main memorial to those who died. I always think of Unit 4 as being the 'front' of Chernobyl, with Unit 1 being at the 'back', purely because most photos you see of the site are taken from the east looking west, and it's at this rear – past Unit 1, on the other side of the turbine hall, near the admin buildings – where I now find myself. From here, I'm granted a sweeping, panoramic view of the entire Chernobyl complex. It's particularly interesting to me because I have somehow never - not even once - seen a photograph taken from this perspective. I take out my camera and pan across the scene, snapping photos to stitch together later. Someone shouts at me to not take a photo of the unremarkable administration buildings nearby (too late), so I turn back towards the memorial. Built into a red stone wall about 5 feet high are 31 black marble plaques, collectively inscribed with the names of the men and woman who died from acute exposure to radiation. In the centre is a red brick arch, from which hangs a black bell. A black slab of marble with the words, "*Life For Life,*" and a symbol of an atom are engraved in the stone. It's understated, but obviously well-tended. I wonder how the families of those countless victims whose names aren't here feel about it; there's no memorial for them.

The next and final stop is Yanov Station, due-west of the plant. On the way, feeling exhausted, I flick through the photos I've taken so far today. The poor weather has spoiled many of them; pity. The bus comes to a halt. Here already? I step out to the sight of two ageing but majestic diesel locomotives, their flanks perpendicular to me as they bask in the fading, late afternoon light. The two engines aren't alone, I realise, as I walk through the nose-to-nose gap between them and emerge onto a line with four sets of rails. All but one are occupied. I peer down the track in both directions; each side's rails stretch in an unbroken line, converging as they touch the sky. Most striking among the smattering of machinery, and standing out amongst its rusted compatriots, is a brand new, bright yellow mobile railway crane. What could that be for? One potential answer sits nearby: a flatbed rail car carrying blackened chunks of chopped wood. Between them sits a stumpy, burgundy liquid container car, which Katie makes a beeline for and scales without a moment's hesitation, soon to be joined by several others.

Her antics inspire me too, so I prop my tripod up against the huge blue diesel I've been photographing and climb aboard. Without bothering to check the cab door's lock – to my eternal regret – I'm up on the roof within seconds. Someone else has had the same idea and stands atop another engine a few hundred feet further down the line. It's at this exact moment, as I gaze down from my vantage point, that the sun punches through the thick clouds above, illuminating the surrounding landscape with warm, saturated colours. The view is perfect. This moment is perfect. Dark clouds; autumn yellows, reds and greens of every shade; heavy, decaying machinery all around me; the glow of the low-hanging sun giving texture to everything in sight.

I hear someone shouting in Ukrainian from the buildings to my left. Someone closer yells at me in English to get down. The engine isn't so abandoned after all! I catch sight of a handful of angry looking men, presumably the drivers, emerge from behind the bus as I scramble back down. Whoops! I grab my tripod and speed-walk to the front of the procession of trains, eager to put some distance between myself and the blue engine on the off chance the owners/drivers come over to give me a kicking. It was disrespectful of me, in hindsight.

I don't want to leave. Now the sun has come out, making me feel warm for the first time today, I'm sad that this wonderful day will be over soon. You know when you're coming to the end of a special novel? That melancholy feeling you get, knowing it'll all soon be over and you almost don't want to continue reading to preserve the moment - I feel like that. I want to spend the evening walking down the tracks alone, listening to the sounds of the Zone, unsure of my destination. I want to return to the plant and go speak to the men and women working on the decommissioning and New Safe Confinement projects. I want to hear their thoughts on the accident, its legacy, what their lives are like in this inhospitable, isolated part of the world, but most of all about their future. I want to spend the night under the stars on the roof of Pripyat's long-abandoned hotel, and contemplate the city by the cold, distant glow of the Moon. Most of all I long to venture inside Chernobyl's damaged Unit 4, to explore its battered corridors and see the reactor for myself, even for a moment. It's not to be. For the final time today Marek summons us back to the bus – the Slavutych train departs soon - but for the first time today I linger. I want to experience this remarkable place for just a few seconds longer.

Back at the station, our route to the platform is blocked by a crowd of workers funnelling themselves through a line of grey full-body radiation scanners. There's no other way around, so I shrug my shoulders, slide my gear underneath the barrier, and step into the human-sized slot. The hand sensor feels cold under my skin as I place my hands and feet on each of the machine's four detectors, hoping this all goes according to plan. The light goes green - I guess that means I'm not dangerously radioactive. One by one, we filter through the scanners and then walk back down the blue-grey platform to the waiting train. I make sure to stay in the vestibule area beside the doors with two others, leaving the seats to Chernobyl's weary workers.

The return journey across the marsh somehow seems even louder and faster than it did this morning, as if every component of the antiquated train is shaking itself loose in a concerted effort to distance itself from Chernobyl. We race past rivers, swamps, deserted track roads and forests without saying a word, all three of us lost in thought. I video some of the journey on my phone to ensure I won't forget exactly what you see when you approach - and leave - the site of one of history's worst man-made disasters.

Back in Slavutych, Dawid, Katie, Danny and I regroup and visit the local store to buy dinner. I approach the manager, a friendly man in his early 30s who speaks a little English, and ask him to teach me to say 'please' and 'thank you' in both Russian and Ukrainian so I can thank the cashiers. He smiles and coaches me. Everything in here looks alien - I can't read any packaging, and don't recognise most of the products. So, in my shy ignorance - and too tired to cook - I buy the only items I do recognise that require no preparation: ice cream and sponge cake.

Photographs & Diagrams

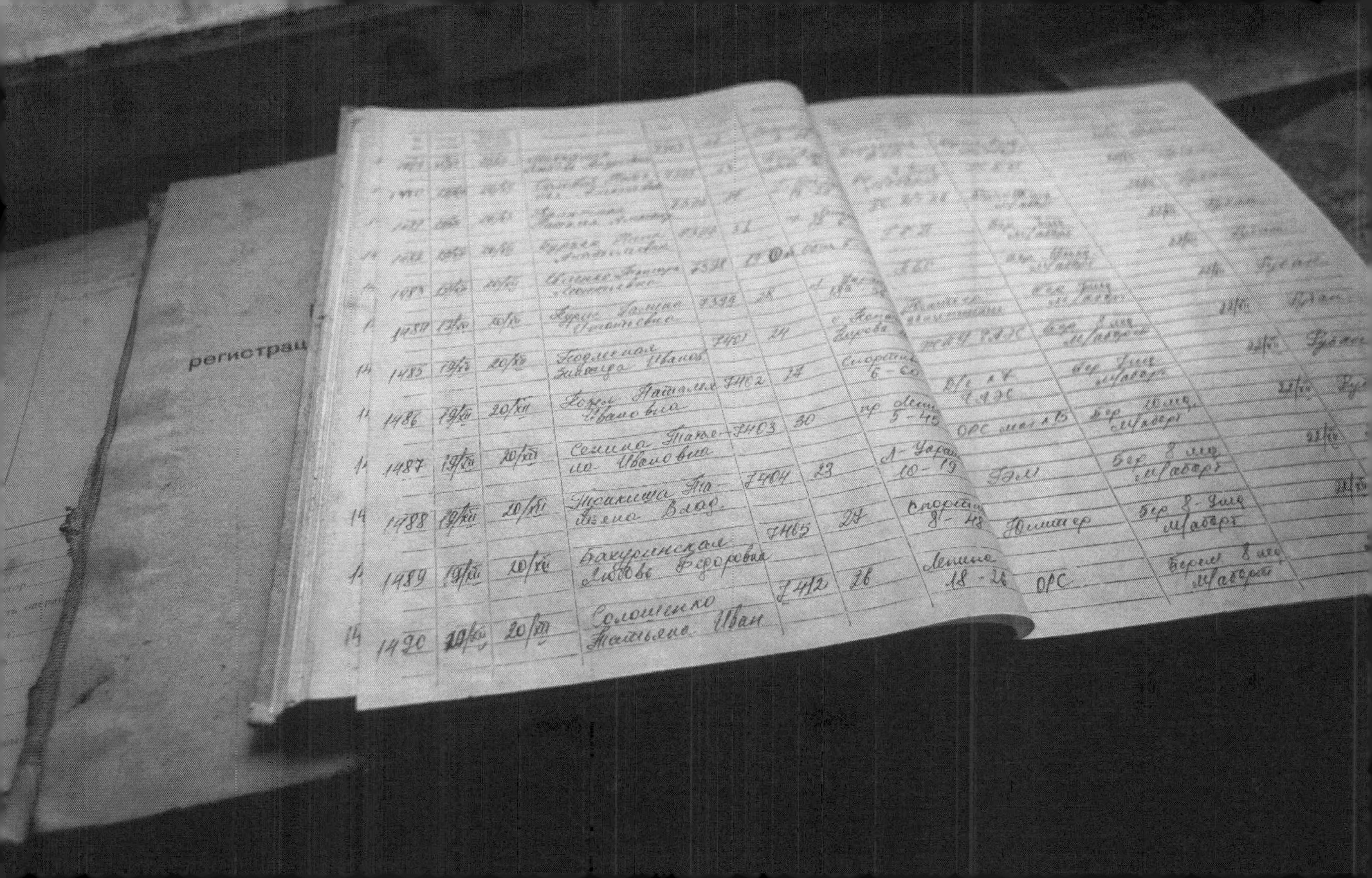
регистрац
1485
1486
1487
1488
1489
1490

Chernobyl Unit 4 Cross-Section

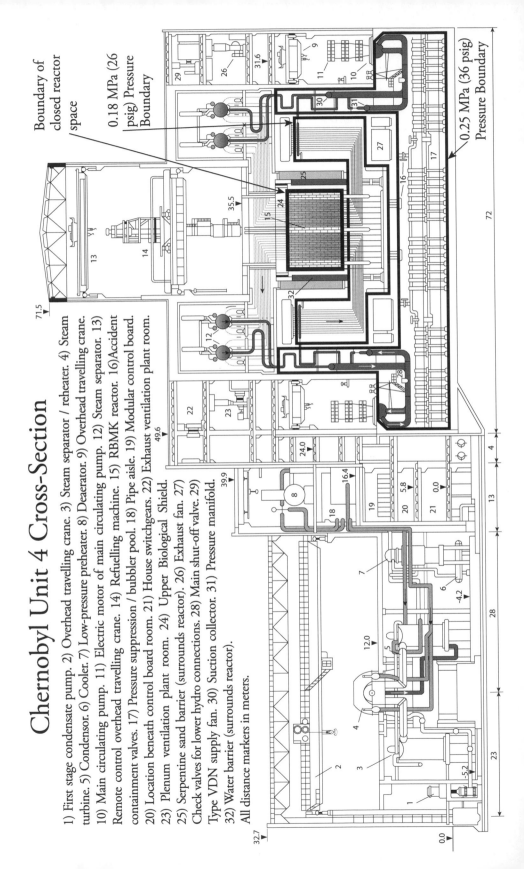

Boundary of closed reactor space

0.18 MPa (26 psig) Pressure Boundary

0.25 MPa (36 psig) Pressure Boundary

1) First stage condensate pump. 2) Overhead travelling crane. 3) Steam separator / reheater. 4) Steam turbine. 5) Condensor. 6) Cooler. 7) Low-pressure preheater. 8) Deaerator. 9) Overhead travelling crane. 10) Main circulating pump. 11) Electric motor of main circulating pump. 12) Steam separator. 13) Remote control overhead travelling crane. 14) Refuelling machine. 15) RBMK reactor. 16)Accident containment valves. 17) Pressure suppression / bubbler pool. 18) Pipe aisle. 19) Modular control board. 20) Location beneath control board room. 21) House switchgears. 22) Exhaust ventilation plant room. 23) Plenum ventilation plant room. 24) Upper Biological Shield. 25) Serpentine sand barrier (surrounds reactor). 26) Exhaust fan. 27) Check valves for lower hydro connections. 28) Main shut-off valve. 29) Type VDN supply fan. 30) Suction collector. 31) Pressure manifold. 32) Water barrier (surrounds reactor). All distance markers in meters.

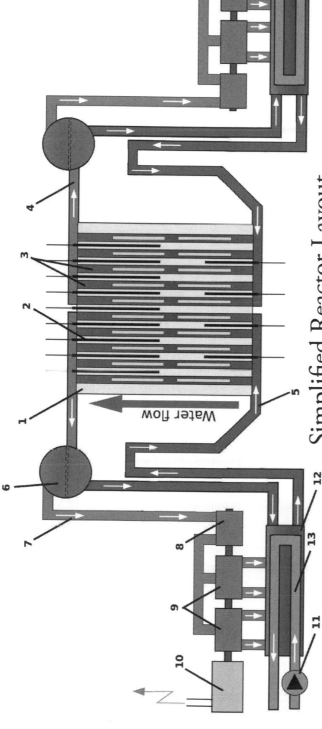

Simplified Reactor Layout

1) Graphite moderated reactor core
2) Control rods
3) Pressure channels with fuel rods
4) Water / steam mixture
5) Water
6) Water / steam separator
7) Steam inlet

8) High-pressure steam turbine
9) Low-pressure steam turbine
10) Generator
11) Main circulating pump
12) Steam condensor
13) Cooling water

Water flow

This picture is my own slight modification of an original image by Stefan Riepl, licensed under the creative commons BY-SA 2.0 terms.

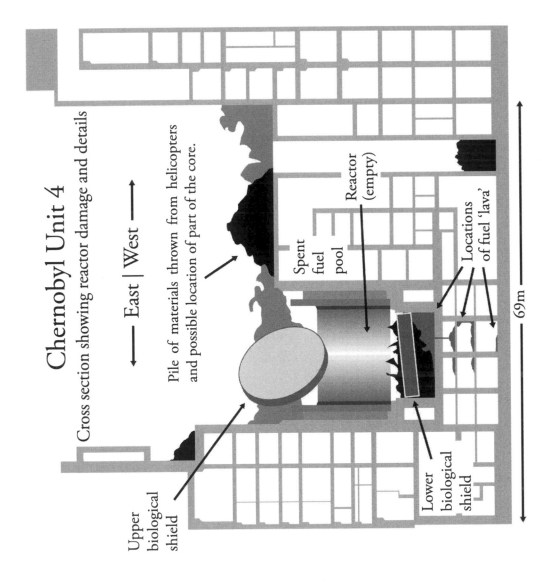

Chernobyl Unit 4

Cross section showing reactor damage and details

◄─── East │ West ───►

Pile of materials thrown from helicopters and possible location of part of the core.

Upper biological shield

Lower biological shield

Spent fuel pool

Reactor (empty)

Locations of fuel 'lava'

69m

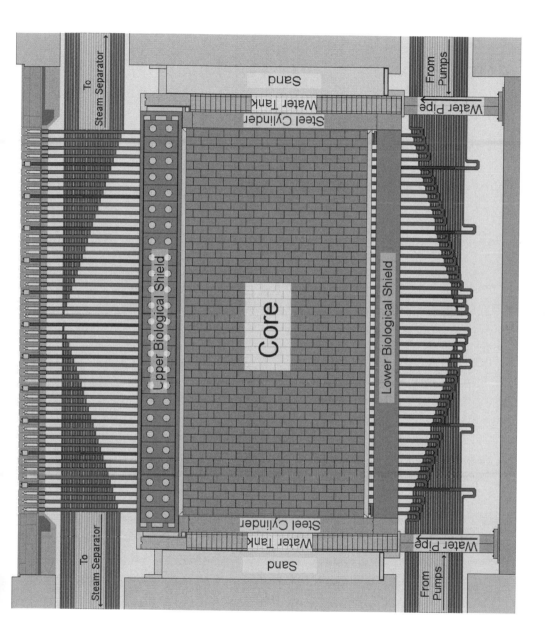

This picture is an original image by 'ChNPP',
licensed under the creative commons BY-SA
3.0 terms.

Alexander Akimov

Leonid Toptunov

Viktor Bryukhanov

Anatoly Dyatlov

CHAPTER SIX

Emergency Response

Plant staff woke and alerted Chernobyl's Manager Viktor Bryukhanov, who returned to the power station by around 2:30am.[150] He ordered for the emergency bunkers to be opened, including the main bunker beneath his administration building, then went straight to his office. On the way, he saw the damaged reactor building and assumed the worst. After failing to make contact with his senior managers from the phone in his office, Bryukhanov called for a meeting in the main bunker.[151] There, he learned there had been a serious accident: a hydrogen build-up had detonated in one of the emergency water tanks, but it was believed that the reactor was intact.[152] Plant workers were preparing a water pump to supply more cooling water to the reactor, and firefighters were tackling blazes on the roof and in the turbine hall; the situation was under control. When

---

[150] Medvedev, Zhores A. *The Legacy of Chernobyl*. Oxford: Basil Blackwell, 1990. Page 47.
[151] Karpan, Nikolaii V. Trial at Chernobyl Disaster. Report. Kiev, 2001. Page 6.
[152] Bryukhanov himself later claimed to have known right from his arrival that the reactor was destroyed, but this contradicts the statements of several others, not to mention his own report to Moscow. See: Medvedev, Grigoriĭ. *Chernobyl Notebook*. Moscow: Novy Mir, 1989. Chapter 4.

questioned on the level of radiation in and around the plant, dosimetrists reported that the only functional radiometer they could find indicated 1,000µR/s - 3.6 roentgens-per-hour. Compared to normal levels this was high, but not immediately life threatening. Bryukhanov and Dyatlov assumed this was an accurate measurement, despite knowing it was the maximum measurement the device could display.[153] In reality, radiation levels reached highs of 8,000,000µR/s - a staggering 30,000 roentgens-per-hour - in some parts of the plant.

Bryukhanov relaxed - 1,000µR/s wasn't so desperate. Local Communist Party officials - men more senior to Bryukhanov within the Party - soon joined him at an imposing table inside the command bunker to discuss evacuation, but feared causing a panic and the possible repercussions if it turned out to be unnecessary.[154] Together, the men chose to assume they faced the best-case scenario. Bryukhanov reported to his superiors in Moscow that the reactor was intact, and thankfully the accident was not as terrible as first feared. They instructed him to write a schedule for getting Unit 4 back up and running, and were assured that the matter would be resolved. Soon after this, staff located a 200R/h radiometer but it, too, went off the scale. Bryukhanov declared the device broken and refused to believe it. Dyatlov and Bryukhanov ignored other staff members sent to fetch readings too, claiming that the men they sent were fools and the devices worthless junk. Within hours, Dyatlov became too sick with acute radiation syndrome to continue working. Despite having now seen for himself the chunks of graphite blocks lying around the plant's grounds, he still didn't accept what had happened.

---

[153] Medvedev, Grigoriï. *Chernobyl Notebook*. Moscow: Novy Mir, 1989. Chapter 3.
[154] Read, Piers Paul. Ablaze: The Story of Chernobyl. London: Secker & Warburg, 1993. Page 110.

The plant's firemen played a fundamental role in preventing the already catastrophic accident from becoming unimaginably worse. Arriving at the scene with his men within minutes, 23-year-old Lieutenant Vladimir Pravik realised straight away that his team was ill equipped and insufficient in numbers to cope with such a widespread and damaging event. He called for backup from all units in Pripyat and the wider Kiev area, then ordered his men to split into teams and concentrate on Unit 3's roof and the turbine hall.[155] The building containing Unit 4 also houses Unit 3 (all four reactors are attached to the same turbine hall), and if the fire were to spread there and take hold, it would have been game over. [156]

Falling reactor debris had ignited everything flammable throughout the site. This, much like the necessity for conducting the run-down test in the first place, was Bryukhanov's responsibility. During construction, the plant's massive roof was supposed to be sealed with a non-combustible material, for obvious reasons. None was readily available in the required quantities, so to proceed on schedule he sourced bitumen instead, of which there was enough in storage.[157] Bitumen is a highly flammable substance, banned from industrial use in the Soviet Union for over a decade (which is perhaps why there was so much of it lying around).[158] The bitumen melted in the intense heat, sticking to the firemen's boots, hindering their mobility and filling their lungs with toxic smoke. Blaming Bryukhanov for this is easy, but he likely had little alternative. There were constant shortages of supplies on a project this large and specialised - the infrastructure to provide for the numerous

---

[155] Medvedev, Zhores A. *The Legacy of Chernobyl*. Oxford: Basil Blackwell, 1990. Page 42.
[156] *The Accident at the Chernobyl Nuclear Power Plant and Its Consequences*. Report. Vienna: USSR State Committee on the Utilisation of Atomic Energy, 1986. Page 25.
[157] Gubaryev, Vladimir.
[158] Medvedev, Zhores A. The Legacy of Chernobyl. Oxford: Basil Blackwell, 1990. Page 43.

nuclear plants being built at the time simply didn't exist. If he had refused to use bitumen, and the plant had fallen further behind schedule, he would have been removed from his position and someone else would have done it instead. Still, I regard using a flammable material to seal the roof of his power station as one Bryukhanov's worst mistakes - there must have been another alternative.

The tragedy of the firemen who arrived to fight the blaze at Chernobyl is that, despite being a nuclear plant fire brigade, many of them didn't seem to comprehend the full dangers of radiation. Those firemen rushing from brigades beyond Chernobyl and Pripyat certainly knew nothing. While there are conflicting accounts on the matter, a number of written testimonies from firemen suggest they hadn't even considered radiation until they became weak and vomited; that a fire was a fire, and they fought it as such. Even then, some thought it was sickness from smoke inhalation and heat. Firemen stationed at nuclear plants in western countries are specially trained and wear specific equipment to help protect them from radiation. In the USSR, nuclear power station firemen wore no special clothing to minimise radiation exposure - not even a basic respirator, only a facemask with an air filter.[159]

One fireman said later: *"We didn't know much about radiation. Even those who worked there had no idea. There was no water left in the trucks. Misha filled a cistern and we aimed the water at the top. Then those boys who died went up to the roof – Vashchik, Kolya and others, and Vladimir Pravik.... They went up the ladder ... and I never saw them again."[160]* Anatoli Zakharov, speaking in 2006, remembers it differently: *"Of course we knew!"* he laughs. *"If we'd followed regulations, we would never have gone near the reactor. But it was a moral*

---

[159] Medvedev, Zhores A. *The Legacy of Chernobyl*. Oxford: Basil Blackwell, 1990. Page 43.
[160] Shcherbak, Yuriy M. *Chernobyl*. Moscow: Yunost, 1987. Page 44.

obligation - our duty. We were like kamikaze."[161] Colonel Telyatnikov was in charge of the second wave of firefighters who arrived 25 minutes after the explosion. *"I cannot tell you now who told me about the radiation,"* he said. *"It was a station worker. They all wore white uniforms. As we were putting out the fire, you had the impression you could see the radiation. First a lot of the substances there were glowing, luminescent, a bit like sparklers. There were flashes of light springing from place to place as if they had been thrown. And there was a kind of gas on the roof where the people were. It was not like smoke. There was smoke, too. But this was a kind of fog. It gave off a peculiar smell."*[162] None of the men he sent to the roof survived, and Colonel Telyatnikov himself died of cancer in 2004, having absorbed hundreds of roentgens fighting the fire. He was 53.

Incredibly, it transpired afterwards that no proper, full fire drill had ever been conducted at the plant. Even the procedure for fighting fire at Chernobyl was almost identical to any other industrial fire, with no regard for the possibility of radiation exposure - so presumptuous were senior figures that nothing could ever go wrong.[163] [164] By 6:35am, when all but the blaze within the reactor core were extinguished, 37 fire crews, comprising 186 firemen in 81 engines, had arrived to battle the flames.[165] A few brave firefighters even ventured inside Unit 4's reactor hall itself and poured water straight into the reactor. The radioactivity was so intense that they received a lethal dose in under a minute. As with most other efforts to cool the reactor over the following days, this only made the situation worse. They were pumping water into a nuclear inferno so hot that

[161] Higginbotham, Adam. *"Chernobyl 20 Years On."* The Guardian. March 26, 2006. Accessed March 09, 2016. http://www.theguardian.com/world/2006/mar/26/nuclear.russia.
[162] Barringer, Felicity. *"One Year After Chernobyl, An Intense Tale Of Survival."* The New York Times (New York), April 6, 1987. Accessed March 09, 2016.
http://www.nytimes.com/1987/04/06/world/one-year-after-chernobyl-a-tense-tale-of-survival.html.
[163] Medvedev, Grigoriĭ. *Chernobyl Notebook.* Moscow: Novy Mir, 1989. Chapter 2.
[164] Medvedev, Zhores A. *The Legacy of Chernobyl.* Oxford: Basil Blackwell, 1990. Page 43.
[165] Ibid. Page 42.

most water either split into a dangerous hydrogen/oxygen mix or instantly evaporated, while any remaining water flooded the basement. Many firemen fell ill in the process, and were rushed to hospital in Pripyat, though it was not well prepared to deal with radiation sickness. Doctors and nurses were also irradiated because the patients they treated were so contaminated that their own bodies had become radioactive.

Initially, there was only one qualified Doctor at the plant, a 28-year-old Pripyat A&E physician named Valentin Belokon, who raced there with no prior warning of a radiation accident after receiving a call from a colleague.[166] He arrived about half an hour after the explosion, but soon discovered the plant's aid station was almost bare.[167] Still, he did his best with what he had and soon noticed a pattern in the symptoms of the surrounding people: headache, swollen glands in the neck, dry throat, vomiting and nausea. Belokon realised what this meant, but selflessly worked for hours to help the stricken plant personnel and firemen, until he too became ill. *"At 6 o'clock [am] I began to feel a tickling in my throat"*, he said afterwards. *"My head hurt. Did I understand the danger? Was I afraid? I understood. I was afraid. But when people see someone in a white coat nearby, it calms them. Like everyone else I had no respirator, no protective clothing... Where was I to get a respirator? I would have grabbed one - but there weren't any. I telephoned the medical station in town: 'Do we have any?' 'No, we don't.' So that was that. Work in an ordinary gauze mask? It wouldn't have helped."*[168] A second physician soon joined him. Dr Varsinian Orlov spent 3 hours in the reactor area helping to stabilise the fallen firefighters, before feeling what he described as, *"a metal taste in the mouth and headache*

[166] Medvedev, Grigoriï. *Chernobyl Notebook*. Moscow: Novy Mir, 1989. Chapter 4.
[167] Mould, Richard F. *Chernobyl - The Real Story*. Oxford, England: Pergamon Press, 1988. Page 167.
[168] Shcherbak, Yuriy M. *Chernobyl*. Moscow: Yunost, 1987.

sickness."[169] Even ambulance drivers ferrying the wounded to Pripyat hospital got sick from the radiation emitted by their passengers.[170]

Chernobyl's third reactor was in a precarious situation of its own. Once Unit 3's Shift Chief Yuri Bagdasarov realised there was no backup water supply to cool the still-operating third reactor, because all water lines from the emergency tanks were connected to its devastated twin, he asked Chief Engineer Nikolai Fomin - who had by now arrived at the plant - for permission to shut it down. Fomin, who struggled to cope during the crisis, forbade it. By 5am, justifiably fearing the worst, Bagdasarov distributed respirators and iodine tablets to his staff to prevent radioactive iodine from building up in the thyroid gland, and then disobeyed his superior's instructions; he shut down Unit 3 himself.[171] Along with the firemen, he prevented the possible destruction of a second reactor. The decision to shut down Units 1 and 2 was not made for a further 16 hours. Fomin, meanwhile, ordered a trusted senior physicist to investigate the state of Unit 4. Like the others before him, his report of the reactor's destruction was ignored and he, too, later died. Time and time again Bryukhanov and Fomin were told that the reactor was completely destroyed, and time and time again they disregarded everyone who warned them.

Captain Sergei Volodin was an Air Force helicopter pilot who often flew a specially equipped Mi-8 transport helicopter around Ukraine. The aircraft was fitted with a dosimeter that Captain Volodin had used in the past to test radiation levels

[169] *"Anatomy of an Accident: A Logistical Nightmare."* The Washington Post (Washington D.C.), October 26, 1986. Accessed March 10, 2016.
https://www.washingtonpost.com/archive/politics/1986/10/26/anatomy-of-an-accident-a-logistical-nightmare/2b1a1238-d27f-45c8-8995-1dba8370c1ac/.
[170] Read, Piers Paul. *Ablaze: The Story of Chernobyl.* London: Secker & Warburg, 1993. Page 85.
[171] Medvedev, Grigoriï. *Chernobyl Notebook.* Moscow: Novy Mir, 1989. Chapter 4.

around Chernobyl out of his own personal curiosity. Prior to the 26th it had never even flickered. On the night of the accident, he and his crew were on standby for the Emergency Rescue shift covering the wider Kiev area, making his the first aircraft to arrive on the scene. As he flew around Pripyat, an Army Major in the rear measured radiation from a personal dosimeter. Neither wore any protective clothing. Volodin's equipment went haywire as he cycled through its measurement ranges: 10, 100, 250, 500 roentgens. All were off the scale. *"Above 500, the equipment - and human beings - aren't supposed to work,"* he remembers. Just as he was seeing his own readings, the Major burst into the cockpit screaming, *"You murderer! You've killed us all!"* The air was emitting 1,500 roentgens-per-hour. *"We'd taken such a high dose,"* the pilot says, *"he thought we were already dead."*[172]

Chernobyl's full complement of morning-shift staff and Unit 5's construction crews had not received word of the accident and arrived for work at 8am that morning, in spite of the surrounding devastation.[173] The Construction Chief sent his crews home at midday because nobody would tell him what was going on, but most of the plant staff remained. All throughout the day of April 26th, firemen and operators continued to pump water into the reactor, succeeding only in flooding more and more of the basement with radioactive water. Bryukhanov gradually came to his senses once he began facing the fact that the reactor was destroyed. The question of evacuating Pripyat was broached soon after the explosion, but he had felt it too momentous a decision to make without very senior backing. He contacted Moscow again and requested permission to evacuate the city, but Communist Party officials, unaware of the full

[172] Higginbotham, Adam. *"Chernobyl 20 Years On."* The Guardian. March 26, 2006. Accessed March 09, 2016. http://www.theguardian.com/world/2006/mar/26/nuclear.russia.
[173] Medvedev, Grigoriï. *Chernobyl Notebook*. Moscow: Novy Mir, 1989. Chapter 4.

extent of the danger – ironic, since Bryukhanov himself had repeatedly assured them the damage was minimal – refused to consider it. An evacuation would cause a panic and spread word of the accident; nobody was to be warned.[174]

A special government commission consisting of Party officials and scientists were on their way to assess the situation, and would arrive over the following 24 hours. The commission's leader was Boris Scherbina, Deputy Chairman of the Council of Ministers of the USSR and a former Minister for Construction in the oil & gas industry. While not a low-level politician, Scherbina was not a member of the Politburo - the Soviet political elite - because nobody in the government realised how serious the accident was at this stage. The most prominent scientific member of the commission was 49-year-old Academician Valerii Legasov. Legasov held a Doctorate in Chemistry and was something of a prodigy, having enjoyed an unprecedented rise within Soviet scientific circles to become the First Deputy-Director of the prestigious I. V. Kurchatov Institute of Atomic Energy. Even though he was not a specialist in nuclear reactors, he was a highly intelligent, experienced and influential figure, both within the Communist Party and the global scientific community.[175]

Saturday the 26th was a blazing hot spring day. Pripyat's 15,000 children – who were particularly vulnerable to radioactive iodine - went off to school (children in the USSR attended school 6 days a week), while the rest of the city's residents went about their normal day. There was even a wedding held that afternoon. All throughout the day, everyone in the area was being silently irradiated. *"Our neighbour ... climbed up to the roof about*

[174] Ibid.
[175] Mould, Richard F. *Chernobyl Record: The Definitive History of the Chernobyl Catastrophe.* Bristol, UK: Institute of Physics Publishing, 2000. Chapter 19, quoting Legasov, Valerii. *"My Duty Is To Tell About This."* Pravda (Moscow), May 20, 1988.

11:00 hours and lay there bathing in trunks to get a sun tan," recalled Gennadiy Petrov, a former plant worker, while speaking with Grigoriĭ Medvedev. *"He came down once to get something to drink, and he said that his tan looked great today, better than ever before. He said his skin immediately gave off a burned smell. And he was in very high spirits, as though he had had too much to drink... Toward the evening, the neighbour who had sunned himself on the roof began to vomit intensely, and they took him off to the medical station."* Another eyewitness reported: *"Word came about an accident and fire at No. 4 unit. But what exactly happened, nobody knew... A group of children from our neighbourhood bicycled over to the bridge near the Yanov station, to get a good view of the damaged reactor unit. We later discovered that this was the most highly radioactive spot in town... They later came down with severe radiation sickness."*[176]

Unsurprisingly, given that the new city existed solely to house Chernobyl's builders and operators, word soon spread that a serious accident had occurred at the plant. *"Different people learned about the accident at different times, but by the evening of 26 April almost everyone knew,"* remembers Lydumila Kharitonova, a senior engineer. *"But still the reaction was calm, since all the stores, schools, and institutions were open. We thought that meant it was not so dangerous. It became more disturbing as evening approached."*[177] That evening, many of Pripyat's families flocked to their balconies and those of their neighbours, to watch the mysterious glow coming from the stricken reactor.[178] As strange as it may sound, the people of Pripyat and the surrounding areas were very fortunate to have excellent weather on the night of the accident and over the following few days. Had it been raining, radioactivity would have poured from the skies and drained into the Dnieper River,

[176] Medvedev, Grigoriĭ. *Chernobyl Notebook*. Moscow: Novy Mir, 1989. Chapter 4
[177] Ibid.
[178] Aleksievich, Svetlana. *Voices from Chernobyl*. Translated by Keith Gessen. Illinois: Dalkey Archive Press, 2005. Page 155.

drastically increasing the number of victims. Instead, most particles stayed high in the air, where their impact was lessened. They were lucky, too, because of the test's timing: it was a spring weekend, when many people were out of town. Those still at home were asleep indoors, protected from the most lethal period of release.

Anyone trying to leave town soon discovered that police had set up roadblocks to stop anyone going into or out of the area. I can see no other justification than this being another measure taken to prevent rumours of the accident from spreading, since at this point only the isolated city's residents and select Communist Party officials knew of it. Had the roadblocks merely prevented people from approaching the site for their own safety it would be fine, but people were unable to leave too. To prevent panic, officials provided no information about what had happened. This, of course, led to frantic speculation and many people tried to escape, avoiding roadblocks by walking out of town through the surrounding forest. Women were seen pushing prams with unprotected babies through the trees. This area later became known as the Red Forest, after its entire population of pine trees turned red and died from exposure caused by the first, deadliest cloud of particles blown from the reactor. It remains one of the most contaminated places on Earth.

By 2pm on the first day, troops from a special Army chemical unit had arrived at Kiev airport and began their journey to Chernobyl, where they took the first accurate measurements of surface-level radioactivity.[179] The readings were extremely high and increasing all the time. By evening, reliable

---

[179] *"Zone of Responsibility: The Chernobyl Heroes."* Pravda (Moscow), December 25, 1986. Quoted by Mould, Richard F. *Chernobyl - The Real Story*. Oxford, England: Pergamon Press, 1988. Page 97.

measurements were finally taken at the Chernobyl power station itself: thousands of roentgens per hour - a lethal dose within minutes. A few months later, radiation would be routinely measured at 240 points throughout the area, but for the time being there were no remote control-machines available for dosimetry, so humans were sent into the radiation fields.[180] Similarly, there were no remote aircraft to take measurements in the atmosphere, so pilots deliberately flew through dangerous plumes to take readings.

Several leading members of the commission boarded a helicopter to view the plant from above and at last confirmed beyond doubt that Chernobyl's reactor had been destroyed. A crisis meeting was held to discuss an appropriate response. None of the politicians understood the ramifications of what had happened, and wasted precious time with their own uninformed suggestions. After much frustrated debate, Legasov and his fellow scientists convinced them that this was not an accident they could sweep under the carpet - it would have significant, lasting, global consequences - nor one they could tackle with conventional firefighting methods. With limited options available, the group agreed their best course of action would be to drop bags of sand - mixed with boron, dolomite and lead to absorb neutrons, absorb the heat, and cool the fire, respectively - from helicopters hovering above the reactor, straight into the core. Tens of thousands of the heavy bags would be required.

Having resisted repeated requests to evacuate Pripyat and the surrounding area by Legasov since his arrival, Scherbina relinquished on the evening of the 26th, and agreed that the population within 10km of the plant should be moved to a safe distance. However, even this decision was tainted. While the

---

180 Mould, Richard F. *Chernobyl - The Real Story*. Oxford, England: Pergamon Press, 1988. Page 113.

scientists favoured immediate compulsory evacuation, Scherbina decided not to inform the city's residents until late the following morning, leaving them unaware of the perils faced by venturing outside for another night and almost no time to prepare for the evacuation. 1,100 buses in a convoy drove overnight from Kiev to transport the evacuees out of the area. Officials forbade residents from leaving in their personal cars out of concern that they would cause traffic jams and prevent a steady departure.

On the morning of the 27th, as the radiation levels in Pripyat peaked, Legasov remarked that, *"it was possible to see mothers pushing children in prams and children playing in the streets."*[181] To ensure as many residents as possible knew what was happening, people were recruited to travel around town, visiting each home with flyers. At 11am, the evacuation announcement was broadcast via radio around the city. It read: *"For the attention of the residents of Pripyat! The City Council informs you that, due to the accident at Chernobyl Power Station in the city of Pripyat, the radioactive conditions in the vicinity are deteriorating. The Communist Party, its officials and the armed forces are taking necessary steps to combat this. Nevertheless, with the view to keep people as safe and healthy as possible, the children being top priority, we need to temporarily evacuate the citizens in the nearest towns of Kiev Oblast* [an Oblast is another word for a State, Province or County, used in some parts of eastern Europe]. *For these reasons, starting from April 27, 1986 at 2pm, each apartment block will be able to have a bus at its disposal, supervised by the police and the city officials. It is highly advisable to take your documents, some vital personal belongings and a certain amount of food, just in case, with you. The senior executives of public and industrial facilities of the city has decided on the list of employees needed to stay in Pripyat to maintain these facilities in a good working*

[181] Mould, Richard F. *Chernobyl Record: The Definitive History of the Chernobyl Catastrophe.* Bristol, UK: Institute of Physics Publishing, 2000. Page 292, quoting Legasov, Valerii. *"My Duty Is To Tell About This."* Pravda (Moscow), May 20, 1988.

order. All the houses will be guarded by the police during the evacuation period. Comrade, leaving your residences temporarily please make sure you have turned off the lights, electrical equipment and water, and shut the windows. Please keep calm and orderly in the process of this short-term evacuation."[182]

It was an incredibly misleading message. *"I knew that the town had been evacuated forever,"* wrote Legasov in his memoirs, two years later, *"but I couldn't find the moral strength to tell it to the people. Besides, if we told them that they were leaving forever, it would take them quite a while to pack their bags. The radiation levels were already very dangerous, so we told them it was a temporary move."*[183] I sympathise with Legasov's plight, but that sounds a lot like an excuse to me. Had he claimed he didn't want people hauling bags stuffed with radioactive heirlooms I may have accepted it, but to say it would have taken time to pack bags, when they had all morning, doesn't ring true. No meaningful public warning of the dangers faced by remaining in Pripyat was given to enable a smooth evacuation, and there was no hint of a longer absence. Had they been informed of the long-term resettlement prospects, families could have packed everything needed to cope with the transition, and those with cars could have filtered out of town overnight. Instead, people were seen laughing and smiling as they boarded the buses, blissfully unaware they would never return to their homes. On the other hand, there were some residents who understood the gravity of what was happening - workers who understood what had occurred at the plant - and they packed accordingly, but those were few and far between. All dogs, cats and other family pets were left behind. Some were locked in their homes, others were set free; a few chased the

---

[182] *"Timeline Of Events | The Chernobyl Gallery."* The Chernobyl Gallery Timeline. 2013. Accessed March 15, 2016. http://chernobylgallery.com/chernobyl-disaster/timeline/.
[183] *"Chernobyl: Valery Legasov's Battle."* TV-Novosti. 2008.

fleeing buses. Despite a couple of isolated incidents where elderly people refused to leave or hid from their rescuers, the actual evacuation was remarkably efficient and took just over two hours.

Moscow ordered the exclusion zone's initial radius of 10km from the plant expanded to 30km - an area of 2,800km² - six days later, after more extensive radiation readings revealed the severity of the contamination. This required those people who had only been moved a short distance to retreat a second time. In yet another attempt to preserve secrecy of the accident, the population of Pripyat and the nearby villages were only displaced up to about 60km, and were deposited with little organisation all over the surrounding villages and towns. There were reports of families being split up, of hosts refusing refugees entry to their homes, and even people given care of children who were not their own. Because they had been instructed to travel light, many evacuees took no money and no ID papers (vital for pretty much everything in the USSR), which caused additional problems down the line. Lots of people were unsurprisingly unsatisfied with remaining so close to the site of the accident, so they made their own way further afield. One helicopter pilot reported later that he, *"could see huge crowds of lightly dressed people, women with children, old people, walking along the road and the roadside in the direction of Kiev."*[184] Later in May, there was a further evacuation from this 60km line for pregnant women and children after radiation levels continued to be dangerous, while towns as far away as 400km were similarly evacuated due to contaminated rainfall. In total, around 116,000 people were

---

[184] Medvedev, Grigoriĭ. *Chernobyl Notebook*. Moscow: Novy Mir, 1989. Chapter 5, quote by G. Petrov.

moved from 170 villages and towns during 1986.[185] After 1986, a further 220,000 people from Ukraine, Russia and Belarus were relocated.[186]

The 129 men and women who were most heavily irradiated - firefighters, plant workers and one female security guard - were flown from Pripyat Hospital to Moscow's famous Hospital Number 6, which specialises in treating radiation related illnesses. By the time they arrived, they were in a grave condition. The patients' own families were forbidden from approaching them because their bodies emitted too much radiation, and existing patients who had occupied the same floor were moved to other parts of the building for protection.[187] Even the staff became reluctant to approach them; *"Many of the doctors and nurses in that hospital, and especially the orderlies, would get sick themselves and die. But we didn't know that then,"* says Lyudmilla Ignatenko, the wife of a deceased fireman, in Svetlana Alexievich's haunting book, Voices From Chernobyl.[188] Her book contains many haunting monologues. Ivan, a firefighter, remembers: *"I woke up in the hospital in Moscow with 40 other fire fighters. At first we joked about radiation. Then we heard that a comrade had begun to bleed from his nose and mouth and his body turned black and he died. That was the end of the laughter."[189]* This may be a reference to Pravik, who was among the very first to die from exposure. When Hospital No. 6 ran out of room, Hospitals No. 7 and then No. 12 made space for the rest of the most seriously irradiated patients. Unfortunately, unlike with No. 6, no information was

[185] Bennet, Burton, Michael Repacholi, and Zhanat Carr. *Health Effects of the Chernobyl Accident and Special Health Programs*. Report. Geneva: World Health Organisation, 2006. Page 2. The initial Soviet estimate was 135,000, but this was later revised down to 116,000.
[186] Ibid. Page 3.
[187] Gale, Robert Peter, and Thomas Hauser. *Chernobyl: The Final Warning*. London: Hamish Hamilton, 1988.
[188] Aleksievich, Svetlana. *Voices from Chernobyl*. Translated by Keith Gessen. Illinois: Dalkey Archive Press, 2005. Page 6.
[189] *"The Liquidators - Chernobyl Children International."* Chernobyl Children International. Accessed March 16, 2016. http://www.chernobyl-international.com/case-study/the-liquidators/.

ever released about the patients who stayed in these other two hospitals.[190]

Lyudmilla Ignatenko recalls the aftermath in distressing detail: *"The doctors kept telling them they'd been poisoned by gas, for some reason. No one said anything about radiation... He started to change - every day I met a brand-new person. The burns started to come to the surface. In his mouth, on his tongue, his cheeks - at first there were little lesions, and then they grew... The other bio-chambers, where our boys were, were being tended to by soldiers because the orderlies on staff refused, they demanded protective clothing. The soldiers carried the sanitary vessels. They wiped the floors down, changed the bedding. They did everything.* [They were soldiers from the same Army chemical division that took the first readings at Chernobyl - A.L.] *But he-he-every day I would hear: Dead. Dead. Tischura is dead. Titenok is dead. Dead. He was producing stool twenty-five to thirty times a day. With blood and mucus. His skin started cracking on his arms and legs. He became covered with boils. When he turned his head, there'd be a clump of hair left on the pillow... At the morgue they said, "Want to see what we'll dress him in?" I did! They dressed him up in formal wear, with his service cap. They couldn't get shoes on him because his feet had swelled up. They had to cut up the formal wear, too, because they couldn't get it on him, there wasn't a whole body to put it on. The last two days in the hospital - pieces of his lungs, of his liver, were coming out of his mouth. He was choking on his internal organs.""[191]*

Two months later she gave birth to his child. The baby girl only lived for 4 hours before dying of a congenital heart defect. She had cirrhosis of the liver after absorbing 28 roentgens from her father, one of the 29 people to die of acute radiation exposure.

---

[190] Gale, Robert Peter, and Thomas Hauser. *Chernobyl: The Final Warning*. London: Hamish Hamilton, 1988.
[191] Aleksievich, Svetlana. *Voices from Chernobyl*. Translated by Keith Gessen. Illinois: Dalkey Archive Press, 2005. Prologue.

The plant operators spent their remaining agonising weeks alive speculating over what had caused the explosion. *"Every day, those who were recovering would gather in the smoking room of* [Hospital] *No. 6, and they were all tortured by one thing: why did the explosion occur?"* recalled V. G. Smagin, Unit 4's morning Shift Chief, who had arrived to relieve Akimov. *"They thought about it and conjectured. They supposed that the explosive mixture of gases could have built up in the coolant drain tank of the emergency control system. A puff could have occurred, and the control rods shot out of the reactor. As a consequence, a prompt-neutron excursion. They also thought about the 'tip' effect of the control rods. If the formation of steam and the tip effect coincided - again a runaway reactor and explosion. At some point, they all gradually came to the idea of a burst of power."*[192]

The event especially tormented Akimov. Depressed and slowly, painfully, inexorably dying in hospital, he felt that he - as the man who pressed the button which lead to the explosion - was responsible, but could not understand why it had gone so wrong. His wife visited in him hospital the day before he died. *"While he could still talk, he kept repeating to his father and mother that he had done everything right,"* she said to Grigoriĭ Medvedev in 'Chernobyl Notebook'. *"This tortured him to the very end. [The last time I saw him], he could no longer speak. But there was pain in his eyes. I knew he was thinking about that damned night, he was re-enacting everything inside himself over and over again, and he could not see that he was to blame. He received a dose of 1,500 roentgens, perhaps even more, and he was doomed. He became blacker and blacker, and on the day he died he was as black as a negro. He was charred all over. He died with his eyes open."*[193] It was May 10th, a beautiful spring day. The others followed him in rapid succession: first the firefighters, then the operators who received the worst exposure. 26-year-old Leonid

---

[192] Medvedev, Grigoriĭ. *Chernobyl Notebook*. Moscow: Novy Mir, 1989. Chapter 7.
[193] Ibid.

Toptunov died on the 14th. Dyatlov spent 6 months in hospital, but survived.[194]

Dr. Orlov, the 41-year-old second physician to arrive at Chernobyl, also spent his final days in Hospital Number 6. *"When I first saw Orlov, he already bore signs of severe radiation sickness,"* recalls Dr. Robert Gale in his book, 'Chernobyl: The Final Warning'. Dr. Gale is an American who worked with Soviet doctors to save the most critical patients at Hospital Number 6. *"Black herpes simplex blisters scarred his face and his gums were raw with a white lacy look, like Queen Anne's lace, caused by candida infection. Then, over several days, the skin peeled away and his gums turned fire-engine red like raw beef. Ulcers spread across his body. The membranes lining his intestines eroded and he suffered bloody diarrhoea. We administered morphine to ease the pain, but even when delirious, he remained in agony. The nature of radiation burns is that they get worse rather than better, because old cells die and young ones are unable to reproduce as a result of the damage. Towards the end, Orlov was barely recognisable and his death several weeks after the disaster was merciful."*[195]

All told, approximately 100,000 people were examined in the days and weeks after the accident, 18,000 of whom required hospitalisation. It took the combined efforts of 1,200 doctors, 900 nurses, 3,000 physicians' assistants and 700 medical students working in shifts to provide round the clock care.[196]

The world remained ignorant of the accident at Chernobyl until the morning of Monday April 28th (it's April 28th as I'm writing this, strangely enough), when a sensor detected elevated radiation levels on engineer Cliff Robinson as he arrived for work at Sweden's Forsmark Nuclear Power Plant, over 1,000

---

[194] Gale, Robert Peter, and Thomas Hauser. *Chernobyl: The Final Warning*. London: Hamish Hamilton, 1988.
[195] Gale, Robert Peter., and Thomas Hauser. *Chernobyl: The Final Warning*. London: Hamish Hamilton, 1988. Page 58.
[196] Ibid. Page 123.

kilometers away. *"My first thought was that a war had broken out and that somebody had blown up a nuclear bomb,"* says Robinson. *"It was a frightening experience, and of course we could not rule out that something had happened at Forsmark."*[197] After a partial evacuation of the plant's 600 staff, those that remained urgently tried to locate the source of what they assumed was a leak somewhere on site. It became apparent from isotopes present in the air that the source was not a nuclear bomb, as was feared, but a reactor. The Swedish Institute of Meteorology and Hydrology analysed the trajectory of the radioactive particles in the atmosphere, which indicated that they were emanating from the southeast: The Soviet Union. Sweden's Ambassador in Moscow telephoned the Soviet State Committee for the Use of Atomic Energy to ask what was happening, but was told they had no information for him. Further inquiries were made to other Ministries, but again the Soviet government claimed they had heard nothing about any accident. By the evening, monitoring stations in Finland and Norway had also detected the high radiation contents in the air.[198]

The cat was out of the bag, leaving the USSR's leadership with little choice but to begrudgingly admit to the world that an accident had taken place. A brief, equivocate announcement on Radio Moscow revealed little: *"An accident has occurred at the Chernobyl nuclear power plant. One of the atomic reactors has been damaged. Measures are being taken to liquidate the consequences of the accident. Those affected are being given aid and a Government commission of inquiry has been created."* The refusal to divulge any further details, other than a then-accurate but still disbelieved fatality count of two, caused rampant speculation in the Western world. United

[197] *"Chernobyl Haunts Engineer Who Alerted World."* CNN. April 26, 1996. Accessed March 16, 2016. http://edition.cnn.com/WORLD/9604/26/chernobyl/230pm/index2.html.
[198] Jensen, Mikael, and Lindhé, John-Christer. *Monitoring the Fallout.* Report. Vienna: International Atomic Energy Agency, 1986. Pages 30-32 of the August 1986 *"IAEA Bulletin."*

Press International printed a widely re-quoted figure of 2,000 dead from a dubious Kiev source who claimed to be close to rescue workers in the city: *"Eighty people died immediately and some 2,000 people died on the way to hospitals."*[199] The New York Post, meanwhile, decided to one-up the hysteria by printing the ludicrous and provocative May 2nd headline, *"Mass Grave for 15,000 N-Victims."* [200]

With Pripyat's population out of harm's way, concentration returned to extinguishing the reactor fire and preventing any further release of poisonous fission products from the core. Easier said than done, but the commission had the full backing of the Soviet government, meaning any and all resources were at their disposal. Helicopter pilots were withdrawn from the war in Afghanistan and set to work flying constant sorties over Unit 4, dropping sandbags into the molten crater. At first, only three men filled the bags with sand - two Deputy Ministers and Major General Antoshkin of the Air Force. *"We were soon in a sweat,"* recalled Gennadi Shasharin, Deputy Minister of Power and Electrification. *"We worked just the way we were: Meshkov and I in Moscow suits and street shoes, and the General in his dress uniform. All without respirators and dosimeters."*[201] The first few dozen flight crews were soon too sick to continue working, having hovered 200m above the reactor in temperatures up to 200°C, dropping bags one by one by hand, and leaning out of the door to estimate the drop-point. The helicopter's designers soon devised a clever system for dropping around 8 bags per flight by using a

---

[199] Means, Howard. *"How Did Chernobyl Corpse Report Get Into Thousands - And Why?"* Orlando Sentinel (Orlando), May 18, 1986. Accessed March 16, 2016. http://articles.orlandosentinel.com/1986-05-18/news/0220260183_1_factors-chernobyl-nuclear-disaster-thousands.
[200] Rosenstiel, Thomas B. *"Soviet Secrecy Blamed for Exaggerated American Reports on Chernobyl Disaster."* Los Angeles Times (Los Angeles), May 10, 1986. Accessed March 16, 2016. http://articles.latimes.com/1986-05-10/news/mn-4936_1_soviet-union.
[201] Medvedev, Grigoriï. *Chernobyl Notebook.* Moscow: Novy Mir, 1989. Chapter 5.

net hanging from beneath the fuselage, allowing them to release the whole payload via a lever in the cockpit.[202]

Sandbags caused the fire's temperature to drop straight away, but radioactive particles in the air increased sharply because more and more dust and debris was kicked into the air from the heavy falling bags' impact. After the first day, Major General Antoshkin proudly told Shcherbina that 150 tons had been dropped into the reactor. He responded, *"150 tons of sand for a reactor like that is like a BB shot to an elephant."*[203] The General, taken aback, arranged for far more soldiers and pilots to be brought to the Exclusion Zone. These young pilots each flew many times over the reactor, and soon took to placing lead plates underneath their cockpit seats to minimise radiation exposure. Despite their homemade preventative measures, many pilots were fatally contaminated and died.

On April 28th, helicopters dumped 300 tons of sand into the reactor. On the 29th: 750 tons; on the 30th: 1,500 tons; on May 1st, May Day, a popular annual holiday in the Soviet Union: 1,900. In total, around 5,000 tons of materials fell into the reactor. By the evening of the 1st, the daily number was ordered to be halved, as there was a growing fear that the foundations wouldn't hold under the strain of so much additional weight.[204] If that happened, everything could collapse into the large pressure suppression pool (a water reservoir for the emergency cooling pumps, which doubles as a pressure suppression system, capable of condensing steam in case of a broken steam pipe) below. This, in turn, could trigger a steam explosion that, some Soviet physicists calculated, could vaporise the fuel in the three other reactors, flatten 200 square kilometers, contaminate a

---

[202] Medvedev, Zhores A. *The Legacy of Chernobyl*. Oxford: Basil Blackwell, 1990. Page 56.
[203] Medvedev, Grigoriĭ. *Chernobyl Notebook*. Moscow: Novy Mir, 1989. Chapter 5.
[204] Ibid.

water supply used by 30 million people, and render northern Ukraine and southern Belarus uninhabitable.[205]

Putting out the fires around the plant had been an important first step towards bringing the situation under control, but the danger was far from over. It is now known that almost none of the neutron-absorbing boron mix in the sandbags made it into the core. The sandbags had, however, partially sealed the open gap between the slanted Upper Biological Shield and the reactor wall below. This was causing the fire to increase in temperature due to a reduction in heat exchange between the core and surrounding environment. The fire reached at least 2,250°C (the element ruthenium, which melts at that temperature, was detected in radioactive vapour that escaped the core), confirming that a meltdown was occurring.[206] At the same time, the amount of fission products being dispersed into the atmosphere increased. Legasov's sincere plan to save the plant, born out of a desperate need to do *something*, had succeeded only in making the situation worse.

A meltdown is when the core components (fuel, cladding, control rods etc.) of a reactor get so hot that they melt together and become a kind of radioactive magma. This can burn down through a containment vessel and potentially through the concrete foundations of the reactor building. If the molten core were to breach all containment and burn down to the water table in the earth below, there was a chance of triggering a colossal steam explosion, with results much the same as an explosion in the pressure suppression pool. Interestingly, modern Russian reactors have a safety feature designed specifically to deal with this eventuality: a solid pool of metallic

[205] The Battle of Chernobyl. Directed by Thomas Johnson. Play Film / ICARUS Films, 2006. DVD. Documentary. Quoting Vasilli Nesterenko.
[206] Medvedev, Zhores A. *The Legacy of Chernobyl*. Oxford: Basil Blackwell, 1990. Page 60.

alloy lying beneath the reactor. If a melting core breaches its containment vessel, the pool catches it and liquefies, creating currents that swirl the molten core against water-cooled steel walls to prevent it from burning through the foundations.

Running out of options fast, the government commission in charge of the emergency response began what they termed, *"counting lives."[207]* It was abhorrent but inevitable that many, many lives would be lost during the fight to save Chernobyl, so Legasov, Scherbina and other members of the commission discussed their potential contingency plans in terms of how many people would die while carrying them out.

As I mentioned, the most immediate worry was that the reactor core could burn down through the lower biological shield to the pressure suppression pool below, and from there on to the building foundations. Two things were needed to minimise the risk. First, the pool had to be drained, but its two valves in the basement - which could only be turned by hand - were now submerged under radioactive water from the firemen's failed attempt to extinguish the reactor fire. Second, the commission decided that the earth beneath the reactor building should be frozen with liquid nitrogen to harden the ground, support the foundations and help to cool the superheated core.

On May 6th, three incredibly courageous volunteers in wet suits dove into the flooded basement together.[208] The divers were Alexei Ananenko, a senior reactor mechanical engineer who knew the valves' location, and two colleagues: Valery A Bezpalov, a turbine engineer who would turn the second valve, and Boris Alexandrovich Baranov, a shift supervisor who acted as a backup/rescuer in case of an emergency, and who also carried a flashlight. They were aware of the stakes and what

---

[207] Medvedev, Grigoriĭ. *Chernobyl Notebook*. Moscow: Novy Mir, 1989. Chapter 6.
[208] Ananenko isn't sure of the date, but thinks it was May 6th. I've seen May 4th, 6th and 10th.

radiation levels were like in the basement, but were apparently promised that their families would be well taken care of if they died.[209] *"When the searchlight beam fell on a pipe, we were joyous,"* Ananenko told the Government-controlled news agency TASS, shortly after his return.[210] *"The pipe led to the valves."* Their light failed moments later and the poor men had to feel their way along the pipes in darkness. Once the valves were opened, *"We heard the rush of water out of the tank. And in a few more minutes we were being embraced by the guys."* With the valves open, the pressure suppression pool was drained of its 3,200 tons of water, but all three heroic men were suffering from radiation sickness symptoms even as they emerged from the water, and each soon succumbed. Or so the tale goes.[211] [212]

So what really happened, and what became of them? The basement entry, while dangerous, wasn't quite as dramatic as modern myth would have you believe. The pressure suppression pool drainage valves couldn't be reached because most watertight basement corridors and surrounding rooms were full of water. The solution required a team of highly trained firemen wearing respirators and rubber suits to charge their fire engines and the Chemical Troops' protective, armoured vehicles into a loading bay beneath the reactor. There, they placed four special, ultra-long hoses into the water before retreating to the safety of Bryukhanov's bunker beneath the administration building. After three hours of almost zero water movement, the dejected firemen came to the crushing realisation that one of the armoured vehicles must have driven over and severed their

[209] A house / car etc. There is a reasonable chance this is exaggerated folklore, but it seems to be true.
[210] Shanker, Thom. *"Soviet Toll Will Rise: U.S. Doctor."* Chicago Tribune (Chicago), May 16, 1986. Accessed March 19, 2016. http://articles.chicagotribune.com/1986-05-16/news/8602040283_1_radiation-reactor-plant-workers.
[211] Medvedev, Grigoriĭ. *Chernobyl Notebook.* Moscow: Novy Mir, 1989. Chapter 6.
[212] *RBMK Nuclear Power Plants: Generic Safety Issues.* Report. Vienna: International Atomic Energy Agency, 1996. ISSN 1025-2754. Page 24.

hoses. A fresh team brought twenty new hoses and re-entered the reactor building. They emerged an hour later, feeling exhausted and nauseous but triumphant; the replacement hoses were in place, the remaining radioactive water could finally be drained.[213]

Some water remained after the firemen's draining mission, up to knee-height in most areas, but the route was passable. Runners took the first readings in several parts of the basement. There are one or two accounts from reliable sources that mention several others venturing into the basement, but their role is unclear. They may have been the aforementioned reconnaissance missions. It so happened that the firemen drained the basement as Ananenko and his two colleagues came on shift. Baranov was the most senior shift manager, so it was he who decided that Ananenko and Bezpalov should turn the valves, and he would accompany them as their observer/rescuer. The men entered the basement in wetsuits, radioactive water up to their knees in places, in a corridor stuffed with myriad pipes and valves. Each man carried two dosimeters: one strapped to the chest, the other near the ankle. When they entered the main basement corridor, Baranov remained near the entrance while Ananenko followed the pipe he believed to be for the pool. He was correct. His fears that he would not find the correct valve in a darkened maze of concrete and metal proved unfounded, so too was his concern that the valves would be jammed. The water drained away and the men returned to the light.

Folklore varies from death within hours to weeks to months, but TASS - the original source from the time - didn't actually mention any health effects in their initial report. We know probably suffered some ill health, mainly because of the

<hr />

[213] Read, Piers Paul. *Ablaze: The Story of Chernobyl*. London: Secker & Warburg, 1993. Page 179-182.

nature of what they did, but also because of the general radiation situation at the plant as a whole. At the very least, all three were still alive on May 16th, when they were mentioned as being modest about their achievement.[214]

Alexei Ananenko is alive and well. He still works in the nuclear industry, and is involved with the activities at Chernobyl. I spoke with him in March 2016, albeit briefly. There is a patient named Baranov mentioned in Dr. Gale's book, who is said to have died weeks after exposure. However, this is electrician Anatoly Ivanovich Baranov, who died on May 20th from acute radiation syndrome.[215] Boris Baranov died of a heart attack in 2005, aged 65.[216] [217] As for Bespalov, there is very little mention of him. The only indication of whether he's still alive is in a comments thread on a 2015 Russian blog post, where a user rejects the notion that the men died. Clearly not a concrete source, but it's the only information I have. A rough translation reads: *"I do not know where this information is from. Judging by the names and positions, Boris Baranov continued to work at Chernobyl and died in the two-thousands. Valery Bespalov retired from Chernobyl a few years ago, and a year ago he was alive for sure."*[218] [219] Ananenko made a brief but powerful mention of him in his description of their ordeal. *"Trying to recover those distant events, I called my friend Valery Bespalov and he told me about an episode that I do not remember, but*

---

[214] Zhukovsky, Vladimir, Vladimir Itkin, and Leo Chernenko. *"Chernobyl: The Courage to Address."* TASS (Moscow), May 16, 1986. Accessed March 19, 2016. http://www.myslenedrevo.com.ua/uk/Sci/HistSources/Chornobyl/1986/05/16/ChernobylAdresMuz hestva.html.

[215] *"Heroes - Liquidators."* Чорнобильська АЕС. Accessed March 20, 2016. http://chnpp.gov.ua/en/component/content/article?id=82.

[216] Falcon, Vladimir. *"In Memory of a Friend."* Post Chernobyl. April 5, 2005. Accessed March 23, 2016. http://www.postchernobyl.kiev.ua/pamyati-tovarishha/.

[217] Baranov gave an interview to a local newspaper shortly before his death. Unfortunately, the website went offline in late 2015 and hasn't returned. http://tribuna.com.ua/news/124286.htm was the URL, but it now gives a hosting error. I did see it once in 2013/2014, but didn't copy any information from it, not expecting it to go down, and don't remember anything about its contents.

[218] *"The Second Explosion at Chernobyl."* Второй взрыв на ЧАЭС. Accessed March 20, 2016. http://nnm.me/blogs/atck/vtoroy-vzryv-na-chaes/.

[219] A 'Valeriy Bespalov' is listed as an observer from the World Council of Nuclear Workers at a 2004 IAEA General Conference, but I have been unable to find any further information about him.

which very well characterizes the then-situation at the plant. According to him, when we were on the way to the [basement] corridor, Baranov approached the entrance to [a corridor beneath the reactor]. He stopped, pushed the telescopic handle on the DP-5 radiometer to its full-length and stuck the sensor into the corridor. 'I looked over my shoulder at Baranov's readings,' recalls Valery. 'The device went off the scale on all sub-bands. Then followed a short command: "move very quickly." Racing across the dangerous space, I could not resist. I looked back and saw a giant black lump, a fragment of the exploded reactor [fuel], mixed with concrete grit... In the mouth, there was a familiar metallic taste...'[220]

That all three survived for so long after the event is quite a revelation, as the story of the divers who sacrificed their lives to save the plant is one of the most well-known legends to emerge from Chernobyl. Every English-language book, documentary and website I've ever seen that mentions them says they died.

On the same day, oil drilling equipment was set-up on the grounds and prepped to begin injecting liquid nitrogen into the earth beneath the foundations, but the requested nitrogen had been delayed by over 24 hours. Unsatisfied with the delay, the Deputy Chairman of the USSR's Council of Ministers, Ivan Silayev, telephoned Bryukhanov and told him, *'Find the nitrogen or you'll be shot.'*[221] He found it: the tanker frightened drivers were refusing to approach the area, but some military persuasion soon had them moving again, and the nitrogen began pumping before dawn.

Around this time, two senior figures from the International Atomic Energy Agency were invited to the plant. The Agency's Swedish Director General Hans Blix and American Morris

---

[220] Ananenko, Alexei. *"Exposing the Myths of Chernobyl."* Union of Chernobyl. Accessed March 23, 2016. http://www.souzchernobyl.org/?id=2440. Cannot find an exact date for when this was written.

[221] Read, Piers Paul. *Ablaze: The Story of Chernobyl.* London: Secker & Warburg, 1993. Page 185. I use this quote with the caveat that Mr Read does not include a list of references in his book, so I don't know where he found it.

Rosen, head of its Nuclear Safety Department, flew to the site to speak with officials about the accident and the measures being undertaken to limit its consequences. Upon their return, they were questioned by correspondents from Germany's Der Spiegel news magazine, to whom they gave unhelpful, blunt responses. *"Can you tell whether the Soviet reactors are safer, or less safe, than the reactors in the West?" "They are a different type,"* replied Rosen. *"How strong was the radiation intensity?" "We did not ask."*[222]

On May 10th, the temperature and radioactive emissions from inside the reactor started to fall. By the 11th, days after the water finished draining, a team of technicians ventured into the sub-levels of the plant, bored a hole through a wall below the core and poked a radiometer through. It confirmed their worst fears: the molten core had cracked the reactor's concrete foundations and at least partially poured into the basement. There was now next to nothing stopping it from breaking through the foundations of the building itself and reaching the water table below. A better and more permanent solution than injecting liquid nitrogen from the surface was required.

The very next day, delegates from Moscow visited mining towns around the USSR to recruit miners for an operation to cool the ground beneath the destroyed reactor. They were bussed to Chernobyl and began work on the 13th. One miner described the plan: *"Our mission was this: dig a 150-meter tunnel, from the third block to the fourth. Then dig a room 30 meters long and 30 meters wide* [and 2 meters tall] *to hold a refrigeration device for cooling down the reactor."*[223] Scientists worried that pneumatic drills would stress the building's fragile foundations, so the miners were ordered to dig their tunnel by hand. To limit exposure, they dug down 12

[222] Von Franke, K., and H.-P. Martin. *"Das Ist Rin Trauriger Anblick."* Der Spiegel (Hamburg), May 19, 1986. http://www.spiegel.de/spiegel/print/d-13518480.html. (In German.)
[223] *The Battle of Chernobyl.* Directed by Thomas Johnson. Play Film / ICARUS Films, 2006. DVD. Documentary.

meters before heading towards Unit 4. The project took one month and four days, with miners digging 24 hours a day - in a normal mine, that distance would have taken three times as long. Due to the nature of the dig, it wasn't possible to install ventilation holes, so there was a lack of oxygen and the temperature reached highs of 30°C.

Radiation levels inside the tunnel were around 1 roentgen per hour, but because the work was so cramped and demanding, the miners dug without any protective gear - not even their respirators, which became damp and useless within minutes. At the tunnel entrance, radiation reached highs of 300 roentgens-per-hour. The miners were never warned of the true extent of the danger, and every single one of them received a significant dose. Vladimir Amelkov, a miner who participated in the operation, said years later, *"Someone had to go and do it. Us or someone else. We did our duty. Should we have done it? It's too late to judge. I don't regret anything."*[224] The miners achieved their goal of digging a room beneath Unit 4, but the refrigeration machinery was never installed because the core began to cool on its own. Instead, the space was filled with heat resistant concrete. While no official studies have ever been published, it's estimated that one-quarter of the miners - who were all between 20 and 30 - died before they reached the age of 40.[225] *"The miners died for nothing,"* laments Veniamin Prianichnikov, chief of the plant's training programmes. *"Everything we did was a waste of time."*[226]

---

[224] Ibid.
[225] Ibid.
[226] Higginbotham, Adam. *"Chernobyl 20 Years On."* The Guardian. March 26, 2006. http://www.theguardian.com/world/2006/mar/26/nuclear.russia.

CHAPTER SEVEN

Radiation

My alarm interrupts an unbroken eight-hour sleep. Feels more like two. Staggering out of bed, I gather my discarded clothes from the floor, rub the sleep from my eyes, then amble through to our cramped kitchen for a wake-up mug of sweet tea. We head out early; today we explore Pripyat.

I deliberately pack light when travelling. It's a practical decision, I don't want to waste time worrying about suitcases going missing, nor dragging unnecessary weight around with me. Danny, for example, brought an armful of huge photography books with him. Two sets of clothes, a toothbrush and plenty of deodorant will suffice if I'm only away for a couple of days. I concede that it's a little unpleasant, but I hate carrying more than one bag (the only exception is my tripod in its case). Naturally, I counter this by taking far too much superfluous camera gear, and lenses weigh more than socks! More lenses than I'll ever use; more batteries than my numerous memory cards could make use of; battery chargers for my phone, camera and laptop; card readers (two, in case one breaks); cables for everything under the sun (just in case both card readers break); lens hoods; a wide variety of cleaning utensils; tripod attachment for my phone (for video recording - I've never used it), and a menagerie of other random bits and pieces. The predictable result is that the weight and space saved by my lack of clothing is more than compensated for, both in weight and bulk, by a ludicrous, tangled mass of camera equipment. I find myself most regretting this philosophy on long, relentless days with few breaks - days like this one.

I'll need as much energy as I can get, so I devour my plate of chicken, cucumber and tomato with all the fervour of a sprinter at the Olympics. After gathering our gear and filing out onto the damp street, we're greeted by a spectacular sunrise - the nicest I've seen for months. Our entire group assembles to watch red bleed across a vivid blue sky, bringing life to the day and splashing slivers of light on nearby puddles and window panes. Like yesterday, tired-looking men and women drift towards the train station in almost absolute silence. It's like a funeral procession, even conversation between ourselves is scant; perhaps everyone feels more serious after the previous 24 hours. I assume the train travels here straight from Chernigov, some 40km due east of Slavutych. It arrives at the platform empty, save for the driver, so it must not stop at any other town or village along the way. We board - I stand - and are soon rattling past miles and miles of cold, quiet swamps and marshland in every direction. It's October, so no flowers are in bloom, but the landscape beyond my condensation-smeared window is so bleak I can't imagine bright colours ever trespassing here. Despite this, the land of northern Ukraine is some of the most fertile in Europe, so the view must look very different during spring.

Upon our arrival, we endure a bumpy bus ride to our first stop of the day: a barren, muddy waste-burial ground 10km east-southeast of the plant, called Buriakivka. It was the principal site used by Liquidators to bury low-level radioactive waste in 1986, such as buildings, household items and an assortment of vehicles. There are 30 covered ditches in two rows of 15, each roughly 150x50 meters in size and containing 22,000m³ of material.[227] Only one remains empty, the rest appear as grassy mounds in the landscape, and the vehicle graveyard in the site's southeast corner is where I now find myself. *"We only stop five minutes here, this area is very radioactive,"* Marek announces through our translator, his sombre face staring at each of us. *"When I say five minutes, I mean five minutes. Touch nothing. When I shout 'time up', you run - not walk - back to the bus."*

My heart sinks. It looks like there are hundreds of vehicles here, all lined up in organised rows across a vast open space. Where do I even start? First, I spot the armoured personnel carriers that ferried soldiers around Chernobyl, of the same type used by the chemical troops. Next, the bulldozers I've seen in documentaries and Igor Kostin's photos that dug up villages in the exclusion zone - too contaminated to save. I race around without pausing to consider composition in my photos, nor even really looking at my subjects for more than a few seconds. I can spend all the time in the world studying them later. Snap, run, snap, run, snap. Endless nondescript, olive-drab trucks; the occasional gutted bus; tankers; trailers; sections of airframes; fire engines, their red paint almost indistinguishable from the rust. How many of their crews are still breathing?

Surprise! I'm ecstatic to discover part of an STR-1 remote-controlled Moon vehicle, used to push graphite and nuclear fuel

[227] Riabtsev, Volodymyr, and Oleg Nasvit. *Remediation of Chernobyl Site and Actual Status of Sarcophagus*. Report. National Security and Defence Council of Ukraine, 2010.

off the roof of Unit 4, nestled in between two trucks. It's smaller than I expected, with its white/silver paint and chunky metal wheels standing out among the greens, browns and deflated tyres. I stop to take a proper look. When I point it out to a nearby photographer, he stares at me, confused. He doesn't understand its significance and presumably doesn't even know what he's looking at - a pile of junk. There's something almost mythical to me about that roof, it sounds like a legend told around campfires. The radiation was so high that even this robot - designed for use in space, the harshest environment known to man - succumbed, followed by a desperate sacrifice when humans replaced it. Too soon, time's up, and I haven't even covered half of the vehicles here. Some distance from me, I can see pieces of helicopters squeezed among other weird and wonderful pieces of history, but there's no time to photograph them. One day.

Danny, Katie, Dawid and I are all seasoned urban explorers, with years of experience. I have snuck into and photographed abandoned hospitals, schools, mansions, hotels, castles, various types of mills, power stations, train stations above and below ground, distilleries, churches, entire villages, and - my personal favourite - a former Top Secret, Cold War jet engine testing facility named the National Gas Turbine Establishment (NGTE) Pyestock, hidden within a pine wood west of London. Even with all that experience, I have never been anywhere close to the scale of Pripyat.

Today we have six hours in the city. It's obvious that, as with all of the best places, there's far too much to see and do, and far too much ground to cover in the time available to us. While Pripyat is small compared to most cities, both in population and geography, it's still too expansive for a small group to see everything on foot in a day. Deciding ahead of time where to spend our six valuable hours was essential. With this in mind, my new friends and I sat down together the previous night over a round of tea, to plan which buildings we'd visit today. We have devised an ambitious schedule using the photography books from Danny's suitcase as a reference for what looked most interesting. We discovered afterwards that we were the only ones to take this approach, and saw by far the most out of our group. The rest just wandered aimlessly; some even spent the entire day in a single building.

Hospital Number 126 is the only high-value target southeast of our drop-off point, and the building furthest from the bus, so we head there first to get it out of the way. We pass countless high-rise residential buildings, brightly-coloured wall murals and unusual structures I can't identify. Buildings I'd ordinarily spend an entire day inside are ignored altogether in favour of more promising targets. The first stricken operators and firemen were brought to this hospital on the night of the accident. Akimov, Toptunov, Dyatlov, Perevozchenko, Pravik. Each of them spent time here. I wish I knew which ward they were admitted to, or if any of their medical records remain amongst the thousands of papers scattered in every room. Sadly, I wouldn't recognise their names in Cyrillic even if I saw them.

As I approach the building, its sandy brown, tiled exterior partially camouflaged by the trees' golden leaves, I spy a rusting lithotomy stirrup-chair sitting alone by the entrance. I always find myself wondering how these things ended up where they are. Someone, at some point during the last 25 years, has decided to drag that chair out of a room, down a corridor, through the main lobby, down the steps beyond the doors and dump it here. Why would anybody do that? The firemen's helmets, clothes and boots - still radioactive to this day - are discarded in the pitch black basement, but I don't go down to see them. The damp, claustrophobic space is like a maze, and the most contaminated spot in the city. Even with my torch I'd probably get lost, and the risk of inhaling poisonous dust - far more harmful than skin exposure - is high. As with everything in Pripyat, the hospital building has been looted countless times by selfish visitors over the years. At first, thieves bribed or dodged the soldiers to steal valuable items left behind after the evacuation, though some of them later paid the price when their loot was dangerously radioactive. For the last decade or so, many explorers visiting the area out of curiosity have also, sadly, been stealing trinkets that interest them. Sometimes to sell (unforgivable), other times for safeguarding. I understand the temptation. When a piece of history is discarded on the ground, your first instinct is to pick it up and save it, but you must remind yourself that it isn't yours to take. It's all part of the Chernobyl story - its place is where it lies.

I skip the ground floor and climb the concrete stairwell straight to the top, reasoning it'll probably be a little less desecrated than the lower levels. No such luck, the top floor is wrecked too; unsurprising after such a long time. Broken chairs, doors, boxes, strip lights, cupboards and bed frames lie all around. Most wards are bare bones and peeling paint - nothing but empty rooms, thick with dust. Some, however, reveal treasures. Sealed, finger-sized vials on dusty glass shelving, somehow still contain a clear liquid. Rooms packed full of books, hand-written patient records and administrative paperwork. An operating table, complete with the classic circular overhead surgical lamp. A wall panel with full-colour illustrated drawings demonstrates how to craft a splint.

As at Buriakivka earlier this morning, our crushing time limit is a constant weight on my shoulders - I know I can't stop to appreciate the things I see, forcing me to rush around too fast to absorb anything. Barely any of my photographs are being composed in any meaningful way, this is almost purely documentary - the sights and sounds come first, the images a distant second. I feel like I'm doing an injustice to the men and women who suffered here by running around the place like a child, trying to see as much as I can before time forces my hand. Continuing a frustrating pattern that's to be repeated throughout the day, I leave the hospital completely unsatisfied with my photographs.

Next, we're making our way to the music school via the cinema. By that time, we'll be close to the hotel which forms part of a cluster of highlights in the centre of town, along with the Palace of Culture (aka the Cultural Centre), the famous ferris wheel and the dodgems. Once we've spent our allotted time with all of that we'll head towards a kindergarten, via another medical centre, then onto the swimming pool. To cap the day, we'll pay a visit to the main high school before returning to our starting point. A lot of ground to cover in 6 hours and, tragically, there are a lot of very promising buildings - the Jupiter factory, in particular - that we just don't have enough time for.

To commemorate this occasion, Danny has the idea of using our photographs as the basis for a commemorative 25th anniversary photographic book on the legacy of Chernobyl, which is a wonderful idea. True to his word, the book has since been published.

Walking through Pripyat is an otherworldly experience. It's the tail end of autumn and fallen leaves lie everywhere, like a blanket of gold spread across the tarmac. As we walk down cracked paths narrowed by overgrowth, all I see are varying shades of orangey yellow; pavements and buildings clouded by a whiskey colour serve as a constant reminder of the encroaching winter. It's so peaceful, the only sounds are the wind whispering to wrinkled tree leaves that it's their turn to give up and fall, the faint but ever-present tolling of the bell - the distant pile driver - and my own footsteps. It all fills me with an almost unique, unsettling feeling that's impossible to describe, as if I'm dreaming or walking through an enclosed movie set. Everywhere I turn the illusion persists, only this isn't fake and I am not dreaming - I really am in a dead city. Part of me half expects to turn a corner and discover the buildings are simple wooden facades, with bored film crews loitering just out of view, waiting to be called to set.

I say *almost* unique because I've experienced this feeling one other time, in the pitch black, subterranean testing chamber of 'Cell 3' at Pyestock. This is the facility where the engines for Concord and Britain's Royal Air Force and Navy were developed and tested. When you first enter the building above ground, Cell 3 appears to be a rather empty, inconspicuous building. About 7 meters wide by 30 or 40 long, tall windows stretch from floor to ceiling, walkways hang high up on the walls, a few railings stand in the centre of an unoccupied floor. Compared to everything else at Pyestock, it seems dull. When you approach those railings, however, you realise they're surrounding a pit buried in the floor, in which sits a massive cylinder lying on its side and reaching out of sight in either direction. A section of this cylinder is missing from the upper face, but there's no obvious route down. You find a way - in my case, descending a rickety, 60-year-old wooden ladder, graciously left off to one side by previous explorers - and discover yourself standing inside the great machine.

At one end of Cell 3 are 10 grilled vents in a ring around a large central exhaust, which reaches through the machine to where a jet turbine would have been mounted. On the other side is an impressive, industrial-looking sliding door. It's not original, closer inspection reveals that the door is made from wood. This place was used as the villain's lair location for the unremarkable 2005 action-adventure film Sahara, starring Matthew McConaughey. Behind the door you crouch-walk for 15 meters down a narrow cylindrical tunnel until you enter the rear of Cell 3. This is where I felt like I was dreaming; the space is almost impossible to describe. The end of the tunnel expands into a drum shape until it has a diameter of 5 or 6 meters, the remains of unidentifiable machinery are attached all around its scorched, torch-lit surface. The bottom is flooded with a coppery, cloudy liquid, with miscellaneous debris on the surface making it seem deeper than it probably is. Detached ends of dozens of pipes reach around a central, circular, ribbed *thing* mounted on the wall opposite you - a heat sink of some kind, perhaps - above and behind which you can just make out a large, black hole in the ceiling. It reminds me of the subterranean tunnels the Nebuchadnezzar traverses in The Matrix.

While I won't have time to see anywhere near all of them, Pripyat had all the facilities you would expect of a modest-sized city. In addition to the aforementioned hospital and its nearby clinics, there were 15 kindergartens, 5 schools, a vocational school/college and a school of music and the arts for the children, with 1 expansive park and 35 smaller playgrounds for them to play in. Further entertainment was found at any of the city's 10 gyms; 3 swimming pools; 10 shooting ranges; 2 stadiums; 4 libraries and a cinema, or by reading Pripyat's own newspaper. Retail came in the form of 25 shops including a bookshop; a supermarket and various smaller food stores; a sports shop; a shop selling TVs, radios and other electronics; and a large shopping centre on the city's central square. For down time there were 27 canteens, cafés and restaurants distributed throughout the city.

Each winter the buildings here become more hazardous as rainwater seeps in, freezes, expands, and damages the brickwork. When the ice thaws, water washes away the mortar, leading to collapses. School No. 1 has had two such collapses in the last few years, and it can be assumed many other buildings in Pripyat are in a similarly delicate state due to lack of maintenance. Within another 25 years I'd expect quite a few to have fallen down. I'm surprised it's taken such a relatively brief period of time for nature to begin reclaiming the city.

The '*Prometheus*' cinema, so called because an obsidian-black statue of the Greek Titan Prometheus once stood watch over the entrance during the city's heyday, emerges from a huddle of trees. We peek inside, but it's now just a hollow shell, with little of interest. Pressed for time, we can't linger. Through some more urban jungle appears the entrance to the music school beneath an abstract tile mosaic - not something you see every day. It's a nice effect, one of the less bland exteriors in the city. I imagine it was intended to encourage creative and innovative thinking, an architectural and philosophical addition I admire. Inside the hall, a majestic, lone grand piano sits atop an empty stage. It's tragic that such a magnificent instrument has been left here to rot, and part of me is sad that the piano was never stolen, as infeasible as that would have been. At least it could still be played if it had - now there's only a dull, muffled thud as I tap the ivory-stripped keys. Near the back of the hall, someone's placed a solitary chair facing the stage. It looks out of place in a hall that would've been filled with life; the last of its kind. Upstairs, in a practise room with a worryingly spongy floor, I find another piano, this one in a much worse state. All four legs and several keys are missing, its twisted and broken strings exposed to the air like entrails.

I want to write in a little more detail about radiation sickness - technically known as acute radiation syndrome - because it's important to convey exactly what it does to a human being who's received an extreme dose, like those plant workers who saved Chernobyl. Low amounts of radiation are relatively harmless. We're all exposed to the natural background radiation of cities, planes, phones, even the Earth itself every moment of every day, and this is nothing to worry about. While every person's body reacts differently, the following is a good general indicator of the consequences. It's often stated that radiation has no taste, but those who absorbed the highest doses at Chernobyl all reported a metallic taste in their mouths immediately upon exposure, so it seems that if the dose is high enough to kill you, you'll definitely taste it. It should be noted that a dose high enough to kill you will also make your body so radioactive that you'll be a major risk to everyone around you.

Once exposed, nausea and vomiting will begin almost immediately, and within a short space of time your tongue and eyes will swell, followed by the rest of your body. You'll feel weakened, as if the strength has been drained from you. If you've received a high dose of direct exposure - as in this scenario - your skin will blanche dark red within moments, a phenomenon often called nuclear sunburn. An hour or two after exposure, you'll gain a pounding headache, a fever and diarrhoea, after which you'll go into shock and pass out. After this initial bout of symptoms, there's often a latent period during which you'll start to feel like you're recovering. The nausea will recede, along with some swelling, though other symptoms will remain. This latent period varies in duration from case to case, and of course it depends on the dose, but it can last a few days. It's cruel because it gives you hope, only to then get much, much worse. The vomiting and diarrhoea will return, along with delirium. An unstoppable, excruciating pain seethes through your body, from the skin down to your bones, and you'll bleed from your nose, mouth and rectum. Your hair will fall out; your skin will tear easily, crack and blister, and then slowly turn black. Your bones will rot, forever destroying your ability to create new blood cells. As you near the end, your immune system will completely collapse, your lungs, heart and other internal organs will begin to disintegrate, and you'll cough them up. Your skin will eventually break down entirely, all but guaranteeing infection. One man from Chernobyl reported that when he stood up his skin slipped down off his leg like a sock. At high doses, radiation will change the very fabric of your DNA, turning you quite literally into a person other than the one you were before. And then you'll die, in agony.

CHAPTER EIGHT

Decontaminating the Zone

Once the immediate danger of the reactor fire was over, work began on a gargantuan operation to clean up radioactive dust and debris across the newly established 30km exclusion zone - particularly around Chernobyl itself - and to design and construct a gigantic cover over Unit 4 to isolate it from the surrounding environment. Military and civilian personnel throughout the Soviet Union were drafted in for the operation, where they became known as Liquidators - liquidating the disaster's effects. According to the World Health Organisation, some 240,000 men and women working within the 30km Exclusion Zone were recognised as Liquidators between 1986 and '87. The clean-up operation continued on a relatively large scale until 1990, by which time around 600,000 civilian and military personnel had received special certificates confirming their status as Liquidators.[228]

---

[228] Bennet, Burton, Michael Repacholi, and Zhanat Carr. *Health Effects of the Chernobyl Accident and Special Health Programs*. Report. Geneva: World Health Organisation, 2006.

The scope was vast: Liquidators' decontamination shifts ranged from a few minutes up to 10 hours a day, depending on exposure levels. First, they built one large and several small dams along river banks near the plant, to prevent rainfall from gathering up the radioactive dust and debris and washing it into the country's most vital source of water.[229] This gave them time to collect, remove and bury the same material that had been blown across the surrounding area. This included burying the Red Forest, which could not be burned because it would spread contaminated particles. Efforts to decontaminate the forest had failed because wind and rain would continually re-distribute radioactivity.[230] Russia's biggest transport helicopters flew around the clock dropping a special polymer resin to seal radioactive dust to the ground. This prevented the dust from being kicked up by vehicles and inhaled, giving troops time to dig up the topsoil for extraction and burial. Construction workers laid new roads throughout the zone, allowing vehicles to move around without spreading radioactive particles.[231] At certain distance limits, decontamination points, manned by police, intersected these roads. They came armed with dosimeters and a special cleaning spray to hose down any passing trucks, cars or armoured vehicles. Among the more drastic clean-up measures was bulldozing and burying the most contaminated villages, some of which had to be reburied two or three times.[232] The thousands of buildings that were spared this fate - including the entire city of Pripyat - were painstakingly sprayed clean with chemicals, while new asphalt was laid on the

[229] Mould, Richard F. *Chernobyl Record: The Definitive History of the Chernobyl Catastrophe*. Bristol, UK: Institute of Physics Publishing, 2000. Page 203.
[230] Ibid. Pages 196-197.
[231] Mould, Richard F. *Chernobyl - The Real Story*. Oxford, England: Pergamon Press, 1988. Page 113.
[232] Kostin, Igor F., and Thomas Johnson. *Chernobyl: Confessions of a Reporter*. First ed. New York: Umbrage Editions, 2006. Page 48.

streets. At Chernobyl itself, all the topsoil and roads were replaced. In total, 300,000m³ of earth was dug up and buried in pits, which were then covered over with concrete. The work took months. To make matters worse, each time it rained within 100km of the plant, new spots of heavy contamination appeared, brought down from the radioactive clouds above.

Hunting parties spent weeks scouring the zone and shot all the abandoned family pets, which had begun to roam in packs. It was a necessary evil to avoid the spread of radioactivity, prevent decontamination workers from being attacked, and put the animals out of their misery. A quick death was better than slowly dying of starvation and radiation sickness. *"The first time we came, the dogs were running around near their houses, guarding them, waiting for people to come back"*, recounted Viktor Verzhikovskiy, Chairman of the Khoyniki Society of Volunteer Hunters and Fishermen. *"They were happy to see us, they ran toward our voices. We shot them in the houses, and the barns, in the yards. We'd drag them out onto the street and load them onto the dump truck. It wasn't very nice. They couldn't understand: why are we killing them? They were easy to kill, they were household pets. They didn't fear guns or people."*[233] They didn't all die this way. At the beginning of June, Nikolai Goshchitsky, a visiting engineer from the Beloyarsk nuclear power station, witnessed some which had escaped the bullets. *"[They] crawled, half alive, along the road, in terrible pain. Birds looked as if they had crawled out of water... unable to fly or walk... Cats with dirty fur, as if it had been burnt in places."*[234] Animals that had survived that long were now blind.

Decontamination jobs did not lack consequences. *"We were told not to have children for five years because of our work,"* remembers

---

[233] Aleksievich, Svetlana. *Voices from Chernobyl*. Translated by Keith Gessen. Illinois: Dalkey Archive Press, 2005. Page 98.
[234] Medvedev, Zhores A. *The Legacy of Chernobyl*. Oxford: Basil Blackwell, 1990. Page 173. Quoting Goshchitsky, N. *'Upala zvezda poly'n*. Page 19.

Igor, a conscripted Liquidator who helped evacuate families and strip radioactive topsoil.[235] *"How do you explain that to your wife or girlfriend? Most of us didn't and hoped we'd be all right. We had to remove the top layers of soil and load it up on trucks. I thought the burial dumps would be complicated engineering places, but they were like open pits, not even lined with anything! We lifted out the topsoil in one big roll like a carpet with all the worms and bugs and spiders inside. But you can't skin the whole country; you can't take everything that lives in the earth. We stripped thousands of kilometers, not just of earth but of orchards, houses, schools - everything. At night we drank so hard. Otherwise we couldn't do it."* Drinking became very common amongst the Liquidators, especially because they were lead to believe that vodka helped protect them from the radiation.

One massive project, nicknamed The Wall in the Earth, was an attempt to isolate Chernobyl from the ground water around it. It is described in Zhores Medvedev's 'The Legacy of Chernobyl': *"A waterproof wall would prevent groundwater from the most contaminated area leaking into the Pripyat river and other water sources... The water-resistant layer of clay was located 30m below the surface. A giant trench was dug more than 32m deep* [and 60cm wide] *around the plant site and was filled with special waterproof bentonite concrete and other water-insoluble compounds. It formed a huge waterproof panel with extra drainage control. The area that had to be isolated from the hydrological environment had to extend far beyond the sarcophagus which was to enclose the reactor (a 2 - 3km radius was probably required)."*[236] [237] A similar project is being attempted at Fukushima, although it involves freezing the earth into a wall of ice, rather than filling it with concrete.

---

[235] *"The Liquidators - Chernobyl Children International."* Chernobyl Children International. Accessed March 16, 2016. http://www.chernobyl-international.com/case-study/the-liquidators/.
[236] Medvedev, Zhores A. The Legacy of Chernobyl. Oxford: Basil Blackwell, 1990. Pages 99-100.
[237] Mould, Richard F. Chernobyl - The Real Story. Oxford, England: Pergamon Press, 1988. Page 136.

Liquidators wore inadequate protective clothing throughout the entire clean-up operation, which was of particular concern to those working near Chernobyl itself. They were assigned just three sets of clothing, and were then expected to wear the same three sets for six months. Others exhibited a rather carefree attitude towards their own future health. *"Our protective gear consisted of respirators and gas masks, but no one used them because it was 30°C outside"*, says Ivan Zhykhov in Voices From Chernobyl.[238] *"If you put those on it would kill you. We signed for them, as you would for supplementary ammunition, and then forgot all about it."* In almost every photograph you can find, Liquidators are without their masks - an invisible enemy was not the concern of young men. Grigoriï Medvedev, the nuclear plant inspector who wrote the prize-winning 1991 book 'The Truth About Chernobyl', recalled, *"I saw soldiers and officers picking up graphite with their hands. They had buckets and were collecting it by hand... There was graphite lying around everywhere, even behind the fence next to our car. I opened the door and pushed the radiometer almost onto a graphite block. Two thousand roentgens an hour. I closed the door. There was a smell of ozone, of burning, of dust and of something else. Perhaps this was what burnt human flesh smelt like."*[239] Soldiers handling reactor graphite by hand shows how uninformed people were in the early days of the clean-up operation. It's difficult to imagine any of the men Medvedev saw survived. For the most part, the Liquidators slept in simple tents erected across vast swathes of land. Some who worked close to the reactor were lucky enough to be lodged aboard one of eight plush passenger liners moored 50km down the Pripyat river, which acted as floating hotels for the weary workers.[240] Pripyat's

[238] Aleksievich, Svetlana. Voices from Chernobyl. Translated by Keith Gessen. Illinois: Dalkey Archive Press, 2005. Page 160.
[239] Medvedev, Grigoriï. *Chernobyl Notebook*. Moscow: Novy Mir, 1989. Chapter 6.
[240] Mould, Richard F. *Chernobyl - The Real Story*. Oxford, England: Pergamon Press, 1988. Page 135.

swimming pool, along with several other leisure facilities, was meticulously and repeatedly cleaned to provide entertainment during down time. There are black and white photos of Liquidators swimming in the pool, which must have been a great way to relieve the stress of daily decontamination work.

By the end of 1986, the Liquidators had decontaminated more than 600 villages and towns. Army troops travelling in armoured vehicles washed Kiev's buildings continually throughout May and June, and it became a crime to own a personal dosimeter in the city for more than two years after the accident. The government placed strict controls on the sale of fresh food; open-air stalls were banned. These restrictions lead the Head of the Central Sanitary and Epidemiological Service of the Ukraine to remark that, *"thousands of ice cream, cake and soft drink stalls have vanished from the streets of Kiev."*[241]

May Day festivities were held throughout the country on May 1st. Countless people marched through Kiev's streets celebrating, just as the radiation intensity reached its peak. There had been no public warning; they were all contaminated. Who knows how many people developed health problems from being out on that day and those that followed. On May 15th, far too late, the city of 2.5 million people was evacuated of its children, their mothers, and pregnant women for four months.

It was obvious from the start that Unit 4 at Chernobyl's Unit 4 could not simply be buried along with everything else, and would instead have to be contained within some new structure. Though its bland official name was The Object Shelter, the men and women who built it soon gave the stark metal and concrete structure a more morbid title: The Sarcophagus. One of the largest and most difficult civil

---

[241] Izvestia, May 2, 1986.

engineering tasks in modern history, there had never been such an important building designed and built in such a short time, under such extreme conditions. It was to stand 170 meters long, 66 meters-tall, and envelop the whole of Unit 4. The Sarcophagus needed the strength to withstand Ukrainian weather for an estimated 20 years - time to develop a more permanent solution - and contain the astronomical levels of radiation within. Erecting the enclosure involved a quarter of a million workers, all of whom reached their lifetime maximum dose. In order for the Sarcophagus to be built, the radioactive graphite and reactor fuel first had to be cleared up and buried, so remote control bulldozers were brought in from West Germany, Japan and Russia to dig up the earth. Workers had originally piled up rubble at the base of Unit 4 and poured concrete straight onto it, intending to seal in the radiation, but that didn't last long. *"Geysers are starting to shoot up from the wet concrete. When the liquid falls on the fuel in the pile, there is an atomic excursion or simply a disruption of heat exchange and a rise in temperature. The radiation situation deteriorates sharply"*, reported Vasiliy Kizima, chief of the construction project at the time.[242]

The biggest obstacle to the Sarcophagus' construction were untold thousands of shattered graphite chunks ejected from the reactor core and thrown up onto the roofs of Unit 3 and the shared chimney. They needed to be removed, but radiation levels on top of Units 3 and 4 - which were too unstable to support the weight of a heavy bulldozer - were far higher than any human could survive. The solution was to airlift remote control robots from across Russia, Germany and Japan, including a couple of lightweight, experimental, remote controlled STR-1 robots from the Soviet space program, built to

---

[242] Medvedev, Grigoriĭ. *Chernobyl Notebook*. Moscow: Novy Mir, 1989. Chapter 6.

land on the Moon, and use them to slowly push rubble off the side of the building. Sixty meters below, the bulldozers would gather up any debris and bury it. In an interesting but tragic twist, however, some robots became stuck in the melted bitumen or tangled in the mangled wreckage, while the rest soon succumbed to the radiation.

"At the beginning, we thought that in some extremely contaminated zones we would use robots," wrote Igor Kostin in his 2006 photo-book 'Confessions of a Reporter'. *'We had even sent a very sophisticated German robot to the plant's roof. But he refused to obey - the radioactivity had disturbed even the machines. Afterward, he rolled over to the edge of the roof and threw himself from the top of the plant. One might have thought he had jumped."*[243] Even the big, modern German remote-bulldozers on the ground broke down. Remote machinery pushed 90 tons of radioactive material from the roof, but it wasn't close to enough. On the ground, they were replaced with manned equivalents; their lead-lined cabs a vain effort to protect each occupant. High up on the roof there was no alternative: men would work in an environment so hazardous that even the machines had died. *'The best robots,"* explains a bitter Nikolai Steinberg, Dyatlov's predecessor, and the man who took over as plant manager from Bryukhanov, *"were people."*[244]

10,000 roentgens per hour is enough radiation to kill you in one minute and was by far the highest level of radioactivity faced by any of the Liquidators. They nicknamed themselves Bio-Robots for the occasion. Nobody had ever worked in such conditions - before or since. *"Obviously some people didn't want to go,"* recalls Alexander Fedotov, a former Bio-robot, *"but they had*

---

[243] Kostin, Igor F., and Thomas Johnson. *Chernobyl: Confessions of a Reporter*. First ed. New York: Umbrage Editions, 2006. Page 49.
[244] Higginbotham, Adam. *"Chernobyl 20 Years On."* The Guardian. March 26, 2006. http://www.theguardian.com/world/2006/mar/26/nuclear.russia.

to - they were reservists. They had to go. For me there was no question, I had to do my duty. Who was going to do it for me? Who was going to clear up this disaster and stop the spread of radioactivity all over the world? Somebody had to do it."[245] And so it was. Scientists calculated that people could work on the roof for up to 40 seconds at a time without receiving a fatal dose. During the day, terrified men from all walks of life dashed across the roof, hurled reactor graphite weighing up to 40 - 50kg over the precipice, and ran back inside. They wore hand-sewn, lead-plated suits that could only be used once (the lead absorbed too much radiation) as their only protection. At night, scouts - nicknamed Roof Cats - scampered over the ruined roof with dosimeters, mapping pockets of radiation so their daytime counterparts could avoid the most contaminated spots.[246]

The reality of the imposed 40-second time limit, it seems, was not always adhered to, at least according to former Bio-robot Aleksandr Kudryagin. *"You were supposed to be up there forty, fifty seconds, according to the instructions"*, he says. *"But that was impossible. You needed a few minutes at the least. You had to get there and back, you had to run up and throw the stuff down - one guy would load the wheelbarrow, the others would throw the stuff into the hole there. You threw it and went back, you didn't look down, that wasn't allowed."[247]* To cope with their fears, the men made light of the situation: *'An American robot is on the roof for five minutes, and then it breaks down. The Japanese robot is on the roof for five minutes and then - breaks down. The Russian robot is up there two hours! Then a command comes in over the loudspeaker: "Private Ivanov! In two hours you're welcome to come down*

[245] *The Battle of Chernobyl.* Directed by Thomas Johnson. Play Film / ICARUS Films, 2006. DVD. Documentary.
[246] Kostin, Igor F., and Thomas Johnson. *Chernobyl: Confessions of a Reporter.* First ed. New York: Umbrage Editions, 2006. Pages 70-71.
[247] Aleksievich, Svetlana. *Voices from Chernobyl.* Translated by Keith Gessen. Illinois: Dalkey Archive Press, 2005. Page 190.

and have a cigarette break."[248] While American purpose-designed robots for high-radiation environments did exist, they were never sent to Chernobyl, despite the joke. The United States offered their services, but the Soviet Government declined.

Where normally it would take one man an hour to do a job, on Chernobyl's rooftop it took 60 men. The work took two and a half weeks, and in most cases each man only went up once - though some went up to five times, and the scouts many more times even than that. Only around 10% of the clean up on the roof was accomplished by actual machines. The rest was done by 5,000 men who absorbed a combined 130,000 roentgens, according to Yuri Semiolenko, the Soviet official responsible for decontamination of the plant.[249] Vladimir Shevchenko, a filmmaker from Kiev, died within a year of filming harrowing roof-top footage of the ruined reactor and Bio-robots entirely without protection. His cameras became so radioactive they had to be buried.

Once the roof was cleared, progress on the prefabricated Sarcophagus proceeded with great haste. In all, more than 400,000m³ of concrete and 7,300 tons of steel were used during 206 days of construction, which concluded in late November, 1986. Because engineers could not manually screw on connecting bolts or weld seams in many places, nor resolve instances where the underlying building sagged under the additional weight as sizeable components were laid down, the Sarcophagus has many unintended holes. The sides and roof of the structure merely rest upon steel support beams - which in turn sit atop damaged concrete - so the Sarcophagus was never

---

[248] Ibid. Page 191.
[249] Anderson, Christopher. "*Soviet Official Admits That Robots Couldn't Handle Chernobyl Cleanup.*" The Scientist, January 20, 1990. Accessed March 22, 2016. http://www.the-scientist.com/?articles.view/articleNo/10861/title/Soviet-Official-Admits-That-Robots-Couldn-t-Handle-Chernobyl-Cleanup/.

particularly strong, and suffered leaks right from the start. Still, it wasn't a major problem, as the structure was never intended to form a hermetic seal in the first place - if it had, there would be a dangerous build-up of pressure. Having already released over 400 times the amount of radiation as the Hiroshima bomb, Chernobyl's 740,000m³ of contents are expected to remain radioactive for thousands of years, and contains enough plutonium to kill millions.

Despite the time limit set in place to protect the Bio-robots, a significant percentage of them have died in the years since, and one could safely assume - given the intensity of radiation exposure, however brief - that all of them developed health complications as a direct consequence of subjecting themselves to such extreme doses. For their sacrifice, they received a certificate and a 100 Rouble bonus ($75 US at the time). In theory there was a safe limit to the dosage a person could receive before he or she would be removed from the zone and allowed to return home. In practise, testimony from a range of former Liquidators suggests that health considerations were barely considered at all. *"At the end of our time there we all got the same thing written on our medical cards"*, says Ivan Zhykhov, who worked in the zone as a chemical engineer.[250] *"They multiplied the average radiation by the number of days we were there. And they got that initial average from our tents, not from where we worked."* Helicopter pilot Eduard Korotkov also noticed problems with the way dosage was recorded. *"On my medical card they wrote that I got 21 roentgen, but I'm not sure that's right"*, he says. *"There's a man there with a dosimeter, 10-15 kilometers away from the power station, he measures the background radiation. These measurements would then be multiplied by the number of hours that we flew each day. But I would go from there to the*

---

[250] Aleksievich, Svetlana. *Voices from Chernobyl.* Translated by Keith Gessen. Illinois: Dalkey Archive Press, 2005. Page 164.

reactor, and some days there'd be 80 roentgen, some days 120. Sometimes at night I'd circle over the reactor for two hours."[251] At the same time, some Liquidators - mainly those who volunteered to work in high intensity areas, like the Roof Cats - deliberately didn't record accurate doses, so they could continue their vital work.[252]

An unofficial estimate of Liquidator fatalities from the Chernobyl Union, a union of former Liquidators, is 25,000, with as many as 200,000 crippled because of their exposure to radiation.[253] While this number is likely far higher than the reality, empirical evidence shows a great many are suffering. Twenty years after the accident, one miner described it thus: *"We've all got a bunch of symptoms. Heart, stomach, liver, kidneys, nervous system. Our whole bodies were radically upset by the metabolic changes caused by radiation and chemical exposure."*[254]

The Liquidators' plight could easily cover a book in itself, but for the sake of balance I will limit myself to this chapter. What matters is that they were brave beyond measure. Throughout my research, one thing has been repeated time and again from all of the different sources, and this seems to be the case with the Soviet mentality in general: people were willing to do whatever was necessary. Untold thousands of men and women surrendered their health and lives for us all, and it's an intolerable injustice that the Governments of former Soviet countries have largely forgotten them, after they gave so much.

[251] Ibid. Page 72.
[252] Kostin, Igor F., and Thomas Johnson. *Chernobyl: Confessions of a Reporter*. First ed. New York: Umbrage Editions, 2006. Pages 70-71.
[253] *"Les Chiffres De L'ONU Sur Les Victimes De Tchernobyl Auraient été Sous-estimés."* Le Monde.fr. April 7, 2006. Accessed March 22, 2016. http://www.lemonde.fr/planete/article/2006/04/07/les-chiffres-de-l-onu-sur-les-victimes-de-tchernobyl-auraient-ete-sous-estimes_759215_3244.html. (In French).
[254] *The Battle of Chernobyl*. Directed by Thomas Johnson. Play Film / ICARUS Films, 2006. DVD. Documentary.

CHAPTER NINE

Exploring Pripyat

We're approaching the hotel. I pass some harrowing graffiti: the black silhouettes of children playing, painted on the hotel's restaurant walls. Near one group, someone has written, 'Dead Kids Don't Cry'. Overlooking the square, the hotel 'Polesie' commands one of the city's best views, so we head straight to the roof, climbing past each tantalising floor without a second glance. You can see for miles up here. Chernobyl sits on the horizon behind abandoned homes, while the ferris wheel apex crests a carpet of trees 150 meters away. As the others busy themselves taking photos from the roof, I separate from them and head towards it. Walking outside on my own for the first time, I gaze across the overgrown square with its cracked concrete and recall decades-old photographs of sunny days, pristine rose bushes, parades and smiling faces. It's so lonely here in the present. I'm a solitary sort of person, and have fantasised countless times about how extraordinary it would be to be the last person on Earth, to go anywhere and do anything I wanted with absolute freedom. Post-apocalypse stories, in particular, have always held sway over me. How ironic that, now I'm experiencing a sliver of that imagined existence, it unsettles me so. I stumble upon a circular building that looks like it was a sports facility of some kind - the splintered remains of a boxing ring dominate the centre. I take a photograph and climb back outside, before approaching what is probably the most iconic structure of the Chernobyl accident, other than the power station itself.

It's a strange feeling when you first see something so familiar from photographs with your own eyes, like visiting the Eiffel Tower or the Pyramids, but familiarity does not preclude awe. You know all the major details, the colours and shapes, but there's so much you never noticed before. There's vital context too: you see everything around it, the geography, and distant things you hadn't expected to see from that particular spot. Near the ferris wheel, which was never formally used, as it'd been due to open as part of the May 1st festivities, are the famous dodgems. A dozen plastic and rubber bumper cars sit idle inside the remains of their 10x20m bare steel enclosure; drapes intended to provide shelter from rain are long gone. The exposed metal base is one of the more radioactive parts of the city, but the cars themselves are in decent shape, all things considered. I saw a great photograph of them once and I try to compose my own, but find myself distracted by thoughts of disappointed evacuee children back on May Day, 1986. They had hoped to be where I am now - laughing and smiling as they barrelled into each other.

I suddenly realise I've been on my own for half an hour. I had expected Danny and the others to join me minutes after I left, but there's neither sign nor sound of anyone. Maybe they weren't heading in this direction until later? Did I actually tell anyone where I was going?

I set off back towards the hotel, glancing up at the roof I left them atop, but not seeing any familiar faces gazing over the edge. Perhaps they headed into the sports building from earlier. Down the corridor from the boxing ring, a filthy practice swimming pool lies bone dry. Did Liquidators drain the water, or has it evaporated over time? Regardless, there's nobody here. I stop and listen for distant footsteps - the sound of boots on broken glass is particularly piercing and easy to hear - but nothing breaks the silence. Have they walked off and left me? A square canvas painting, my height, with a celebratory 'CCCP 60' emblazoned in bold white text against the traditional Soviet blood-red, sits propped against a pillar in the building's large entranceway. It turns out that the sports building is the rear of Pripyat's Palace of Culture, one of the city's most recognisable and central landmarks. Palaces of Culture were large Soviet community centres, containing cinemas, theatres, dance halls, swimming pools, gymnasiums and other sports facilities, like the aforementioned boxing ring. By the end of the 1980s, there were well over 125,000 such buildings across the Union. I stride out through the front doors, past tattered chairs, and study the surrounding landscape. Nobody.

After snapping a few squint photos, I return inside. On my way in, not twenty feet from where I'd been standing mere seconds ago, I bump into Dawid photographing the red canvas. Where did he appear from? He smiles and says that Danny and Katie are upstairs somewhere, exploring the building. I bound up the wide staircase into the main exhibition/dance hall, where the entire left side (from my perspective - it's actually the front of the building) is made up of floor-to-ceiling windows, right down the length of the hall. It goes without saying that the glass is long gone, but it must have been quite an impressive room in its day. On a balcony above me, I spot the others taking photos. I'm relieved to see all my companions again. To my right, the brightly-coloured paint of a 10-meter mural depicting a glorious celebration of communism still clings to the concrete; a losing battle.

We regroup and circle the building. At the eastern corner, I pass through a pair of triple-height doors into the back of what appears to be a theatre or concert hall (or both). Before I spoil myself with that, I investigate a cluttered room off to my right, which contains several square painted portraits of Soviet heads of government, each as large as the 'CCCP 60' example I saw earlier. Gorbachev is instantly recognisable, but I'm unfamiliar with the other men. I was expecting to see Lenin or Stalin, but they aren't here - too tempting for looters. Lenin, at least, was here once, standing proud on a banner hanging from the front wall of the building. His depiction must have been stolen in the years since the accident. Photos taken, I eagerly return to backstage in the theatre hall.

The stage area roof is higher than everything else in the Palace of Culture, enabling light racks to hang high up out of view of the crowd. Those same lights are now slumped across the stage. Dozens of metal cables hanging all around me slice through shafts of light from holes in the masonry. I'm half-tempted to go climbing up into the rigging above for a more unusual vantage point, but, upon further consideration, I decide that I quite like my bones intact. All the chairs that were once here are long gone, save for a few discarded dirty and deflated cushions. Strangely, it looks like all of the wall panels have been stolen. Exposed bricks are visible everywhere, and there's a makeshift, obviously hand-made set of rickety wooden scaffolding reaching up to the ceiling in one corner of the room.

It must've been erected by someone who couldn't bring in proper equipment. The only obvious explanation is that they used it to reach up to the tops of the walls to steal whatever was once on them. There's nothing more here, and the others want to check out the ferris wheel, so we head back into the sun.

Having soaked in the atmosphere while my friends take their photographs, then posing for the inevitable group photo, we're on the move once more. Our little ensemble makes a quick stop in a clinic, before conceding that nothing of interest remains - my most notable shot is of an open window, vibrant red leaves creeping past the frame. Next is one of our big targets of the day: a kindergarten named Golden Key, the biggest of 15 junior schools in the city. Photographs of this building litter the internet, and rightly so, it's choc-a-bloc with unique and fascinating imagery. Situated near the centre of the city, not far from the square, Golden Key is surrounded on all sides by residential tower blocks, though I struggle to spot them through the trees. Beneath my feet, more and more toys lie discarded across the ground as we approach. Upon entering, the first thing to catch my eye is a reclining doll perched on an infant's chair, in an otherwise empty classroom. She wears a faded red and white checkered shirt with black pants, but her face and most of her hair is obscured by an old Soviet child-sized rubber gas mask. Some previous photographer, keen to capture an artificially haunting image, has obviously staged the scene, but that takes nothing away from it. It *is* haunting, knowing what happened here.

There's too much to take in. Everywhere I turn there are sights worthy of hours of study - it's overwhelming. I wander without aim through the building, my camera hanging limp from my shoulder. When I force myself to take pictures, they're impossible to compose - there's just so much to see that I can't decide what to focus on. Room after room packed with babies' cots, children's beds (for napping, or was this a boarding school?), tiny desks, chairs, books and gas masks. Toy animals; toy cars; toy dolls; toy bricks; toy instruments; toy cutlery; toy buildings. There are a few obvious highlights, worthy of careful examination. Gathered around a low, white wooden table, barely a foot off the floor, sit a plastic duck and two dolls: one boy, one girl. The vibrant colours of the bright yellow duck and navy blue-suited boy draw your eye, but it's the comparatively drab girl who demands your attention. Her soft-touch silicone face has dried and cracked, then gradually faded to ash grey over 25 years of neglect. The lacy white dress wrapped around her has become dirty over time, and it, too, is now grey. Sepia hair, once neat, is unkempt and sprinkled with delicate spider's web and crumbled flecks of paint, fallen like snow from the ceiling above. The only real colour on the doll is the pale pink plastic body, exposed through tears in her dress, and her piercing sky-blue eyes.

I hate to leave the kindergarten, but time is of the essence and we need to maintain our momentum to complete the list. The city swimming pool, made famous to gamers everywhere in 2007 by appearing in the Pripyat level of Infinity Ward's revolutionary *Call of Duty 4: Modern Warfare*, is next on the agenda. I don't remember when I first saw photographs of it, but it was long before the release of CoD4, and I've recognised images of it over the years since - even before I knew anything about the Chernobyl accident. There's just something about seeing an empty swimming pool that unsettles me; it sticks in your head. Walking via our drop-off point, where the bus still patiently waits for the clock to reach 3pm, I'm glad Danny, Dawid and Katie seem to know where they're going. If I had spent the day exploring Pripyat alone, I may not have found much of anything. One of the things I love about this city is that, because the trees and bushes are so thick, buildings pop out of nowhere, right in front of you. It's happened several times throughout the day, and again upon our approach to the swimming pool. We enter through a single fire-door in a blank wall. Inside it's almost pitch dark, we tread carefully as we inch along the length of the building, through torch-lit changing rooms. At the far corner, we scale a steep set of rusted metal stairs and emerge into the light. Once again, I'm speechless.

How do you document something without preparation that's been photographed so many times, while keeping your own unique look to your image? The answer is you can't, and my images of the pool are identical to everyone else's. At the time of this trip in 2011 I almost exclusively shot everything wide-angle, to squeeze as much content and context into the frame as possible. I wish I could go back there today, because now I would do it in a completely different way, standing in different places, using different lenses at different angles with different camera settings. After crouching beside the pool for a few minutes, I turn around to find that Katie – ever the adventurer - has climbed up and stands atop the higher of the two diving boards, peering over the edge. Must be a better view up there.

I toss my bag and tripod up to the lower platform, and then pull myself up (the lower ladder is long gone) before scaling steps to the high board. The view *is* better. I'm not sure it's quite Olympic size, but the pool is still big; 6 lanes across and a good 15 feet deep. Light pours into the surrounding space from enormous glassless windows. They stretch the entire length of the building, before reaching around the corner and wrapping along most of each end. I suspect some industrious soul has been doing a little cleaning here, as the ceiling panels - no longer hanging from their perch - are also not in the pool, with one or two exceptions. To what end, I wonder. Dawid appears up on the balcony, making me realise I've spent almost my entire time here photographing the pool itself. I need to see the rest of the building, so I dash to a side door, through another changing room, and discover a basketball court, of all things. The polished wooden floorboards have warped and been torn from the floor at one end, it's very picturesque. We're out of time yet again within minutes, it's so incredibly frustrating.

I'm exhausted. Keeping up this pace for hours, plus not having eaten or drunk anything since breakfast, is beginning to take its toll. There's no time for rest, another of Pripyat's many schools awaits. Along the way, we pass through a beautiful natural corridor of trees, standing guard over fallen yellow leaves that stretch far off into the distance. Reminds me of the Yellow Brick Road in Oz.

Pacing through corridors of an artificial kind - concrete, barren, nondescript - we're getting lost, but after a little backtracking we find what we're looking for. Discarded here by looters in search of the tiny pieces of silver in each filter, the cafeteria's entire floor is drowning in an ocean of hundreds - perhaps thousands - of dust-coated gas masks. The remains of a globe break the surface, its European side is shattered and buried.

Only one more building to go - another secondary school - but it's destined to disappoint. We made our target list based on one of Danny's photography books, but it's years old and the school has been gutted in the intervening time, leaving most rooms bare. I photograph a couple of the more interesting ones, then decide to spend the final 20 minutes just soaking in the atmosphere of this amazing place. I head to the roof, joined by Katie, where we contemplate a silence that'll last ten thousand years.

CHAPTER TEN

Complex Expedition

For six months following the accident, as the Sarcophagus was under construction, a team of courageous scientists from the V. I. Kurchatov Institute of Atomic Energy re-entered Unit 4 as part of an investigation aptly named Complex Expedition.[255] *"Everyone was afraid of one thing: an explosion might happen again, because the reactor was out of control,"* recalls Viktor Popov, the nuclear physicist leading the expedition. *"Were conditions inside the reactor such that another catastrophe might occur?"*[256] In what would be considered a suicide mission under any other circumstances, the scientists' first aim was to find out what happened to the nuclear fuel, and then determine if further spontaneous nuclear reactions were possible. They explored the plant's ruined and unpowered sub-levels with flashlights and cotton masks. *"At that time,"* says

---

[255] I have found it almost impossible to find information on the Complex Expedition. Most of the information in the next few paragraphs comes from the BBC Horizon documentary "*Inside Chernobyl's Sarcophagus.*" Given the pedigree of the source and the fact that they conducted numerous interviews with the scientists themselves, I trust that the information presented is accurate.

[256] Horizon - Inside Chernobyl's Sarcophagus. BBC, 1996. VHS. Documentary.

Popov, *"there were no places in [Unit 4] that were not dangerous, not by normal human standards. We entered fields of 100, 200, 250 roentgens an hour... This kind of situation could crop up unexpectedly. You're walking down a corridor and the levels are not too bad; 1 to 5 roentgens per hour. Then you turn a corner and suddenly it's 500 roentgens! You have to turn and run for it."*[257]

After a long and arduous search, the scientists found fuel in December, with the help of remote cameras poking through a long hole drilled into a wall. It was still emitting 10,000 roentgens-per-hour. *"It made us treat it with the utmost respect,"* remembers Yuri Buzulukov, another expedition scientist. *"To approach it meant certain death."*[258] The two-meter-wide mass, which was discovered deep in the basement and quite a lateral distance from the reactor, had poured through a hole in the ceiling and cooled into a dark, glassy substance. They named it 'The Elephant's Foot' due to its wrinkled, circular appearance. The fuel alone couldn't have done this; the glassy effect was a major breakthrough. Samples were required for study, but the miniature robots sent to chip off pieces didn't have sufficient strength to damage the Elephant's Foot. *"After that, a good idea was put forward: if all else failed, we could try firearms,"* laughs Buzulukov. *"First we turned to the Army. The army sent us to the Police. The Police sent us to the KGB, then finally we tried the Police again, who supplied us with an [AK-47 assault rifle]. They lent it to us on the condition that we would use their volunteer, a very nice, charming man who would shoot specific targets which we indicated to him. The next day, without any difficulty, he shot all 30 rounds at the targets that I pointed out to him, with the help of a video camera. He was very calm about it. Eventually, we got samples from the lower section, and it so happened that we shattered the upper part completely, because - to our pleasant surprise - it*

---

[257] Ibid. 08:50.
[258] Ibid 16:50.

consisted of many layers, like the bark of a tree. After each shot, some of the 'bark' would peel off, and we would start on the next layer, and so on. We obtained a huge number of samples, but we spoiled the beauty of the Elephant's Foot."

The team next needed a closer look at the reactor itself, so they brought in oil industry engineers to drill through the reactor's reinforced-concrete containment structure. They finally broke through in the summer of 1988 after 18 months of drilling in harsh conditions. *"There were many theories about what we might find there,"* says Buzulukov, *"but everyone agreed there would be damaged reactor core: graphite blocks interspersed with distorted fuel rods."*[259] The team were in for a surprise; the reactor was empty, its smooth metal interior wall clearly visible. They were shocked. After drilling another hole through the bottom of the reactor, they discovered a few graphite blocks, but the fact remained that the reactor was essentially bare. *"We faced the huge question: 'Where had it gone?'"* laughs Buzulukov.

Since the volume of the Elephant's Foot alone could not account for all the missing fuel, the team turned their attention to the room directly beneath the reactor, where they had already detected enormous levels of heat and radioactivity. Without access to a robot small enough to squeeze down the narrow tunnel they drilled into a wall, the team was forced to improvise. A plastic toy Army tank was bought from a Moscow toy store for 15 Roubles and strapped together with a torch and camera. The makeshift robot's images were abysmal, but a vague, gigantic mass could be seen within the room. Lacking proper protective equipment and unable to venture into many areas of the basement, the expedition scientists toiled for a further year to get a better view of the room. When at last they did, they

---

[259] Ibid. 18:30.

found it devastated by the reactor explosion, but still there was no fuel.

In 1991, the stressed and exhausted members of the expedition realised they had no choice but to venture into the remains of Unit 4's reactor hall themselves. The risk of a possible second explosion was too great to ignore. With no money or proper safety clothing, a special group from the wider scientific team - with their white overalls literally taped to their gloves and boots to prevent dust from creeping in, and only a basic disposable mask to protect their lungs - entered the devastated space. After a treacherous climb over graphite fragments blown out of the reactor and shovelled off the roof, they discovered steaming concrete, heated by the fuel beneath it. Closer inspection revealed radioactive lava - an astonishing find. As the team passed through a narrow, shattered corridor adjacent to the reactor base, their torch-lit dosimeter crackling away at a frightening 1,000R/h, one man noticed that the lower biological shield had crushed the wall beneath it. The final missing piece of the puzzle fell into place.

On that fateful morning in April 1986, the explosion that blew off the reactor lid also dislodged special serpentine sand and concrete from within the thick walls surrounding the RBMK. In that same moment, a powerful shock wave forced the entire bottom half of the core assembly - including the lower biological shield - downward by several meters, into the space below. Over the following week, intense heat from the fire and radioactive decay increased until it reached temperatures sufficient to melt the fuel assembly, which poured out and bonded with the sand/concrete mix to form a kind of radioactive lava called corium. This lava then oozed through pipes, ducts and cracks in the damaged structure to the rooms beneath. The Elephant's Foot was one offshoot of this lava, which had cooled into a glassy form. Melted fuel vacating the exposed reactor like this is probably what caused the sudden drop in temperature and emission levels in early May, 1986. A molten core is capable of burning through 30cm of concrete within hours, hence the scramble to prevent this from happening.[260]

In the fuel's diluted state, and without the possibility of contact with water, the scientists concluded there was limited risk of another explosion. By 1996, however, things had changed. Condensation and water had entered the Sarcophagus via its many holes and seeped down into the solidified fuel-lava. It reacted with the uranium within, causing a surge in radioactivity. The Sarcophagus was ten years old at that stage, and an estimated 70% probability of collapse within the subsequent decade meant money was redirected away from research and towards engineering. This dangerous situation ultimately resulted in the Designed Stabilisation Steel Structure mentioned in Chapter 5. Research conducted into the corium has since been insubstantial.

With the Object Shelter construction under way in 1986, global attention turned to the Soviet elite whose task it was to find those responsible for the Chernobyl disaster. Possible culprits included: the plant's control room operators from the Ministry of Power and Electrification, who had caused the accident; scientists from the Kurchatov Institute, who designed the technology used in the reactor; senior designers at the Scientific Research and Design Institute of Power and Technology (NIKIET in Russian), who designed the plant itself; Ministers from the secretive Ministry of Medium Machine Building, who approved a reactor they knew had numerous

---

[260] "*Nuclear Reactor Severe Accident Experiments.*" Argonne National Laboratory. July 28, 2014. http://www.ne.anl.gov/capabilities/rsta/cci/index.shtml.

major flaws (although this was never mentioned in public) despite knowing the potential associated risks; members of the State Committee for Safety in the Atomic Power Industry, who held broad control over nuclear safety.

The matter was debated and decided over two meetings of the Interdepartmental Science and Technology Council on the 2nd and 17th of June, 1986. V. P. Volkov, head of the Kurchatov Institute's RBMK safety research group, supplied information to the Council clarifying that the accident resulted from critical design flaws, but the idea that Soviet reactors were anything less than perfect could never be admitted to the world. The USSR was founded on a belief in science, it had always taken pride in being a technological superpower and there was fear among the Council of a possible public backlash against nuclear power, as had occurred in America after Three Mile Island. No, it was a foregone conclusion who the scapegoats would be: Chernobyl's operators. That's not to say that certain members of staff weren't guilty of negligence - they were, without a doubt - but even their disregard for safety would not have caused an accident of such global magnitude had the RBMK been designed properly in the first place.

There were several high-ranking dismissals. The chairman of the State Committee for Safety in the Atomic Power Industry; the Ministry of Medium Machine Building's First Deputy Minister; the Deputy Minister of Power and Electrification (Gennady Shasharin, who had diligently shovelled sand for the helicopters in his expensive suit back in April, and had later attempted to release a report revealing the true cause of the accident); the principle designer of that model of RBMK from NIKIET.[261] All lost their jobs. Over 65 lesser Communist Party

---

[261] Read, Piers Paul. Ablaze: The Story of Chernobyl. London: Secker & Warburg, 1993. Page 270.

officials and staff members from Chernobyl were either fired or demoted, almost half of whom were also expelled from the Communist Party.[262] I couldn't determine who or why, though there were a few deserters who abandoned their posts in the aftermath of the explosion, so they may contribute to the figure. During August 1986, the KGB arrested six men for their role in the disaster. They were: Plant Manager Viktor Bryukhanov, who spent almost a year in solitary confinement awaiting the trial; Chief Engineer Nikolai Fomin; Deputy-Chief Engineer Anatoly Dyatlov, who wrote the turbine test program; Shift Supervisor Boris Rogozhkin, the man in charge of the night shift on the 26th; Yuri Laushkin, the government's safety inspector at Chernobyl; Manager of the Reactor Workshop Alexandr Kovalenko, who approved the test along with Bryukhanov and Fomin. Their trial was set for March 1987, to allow prosecutors time to gather evidence of exactly what had gone wrong, but it was postponed to July 7th after Fomin attempted suicide in his cell. He smashed his glasses and cut his own wrists with the shards of broken glass, but was discovered and saved by prison staff.[263]

In an improvised courtroom, the deserted town of Chernobyl's own Palace of Culture played host to the last of the USSR's show-trials. Soviet law required that the trial take place near the scene of the crime, and radiation provided a convenient excuse to limit the number of attendees, since access to the zone required special papers. Ostensibly an open trial with journalists and victims' families invited to the opening and closing days, the bulk of the three-week proceedings took place in secret, behind closed doors. Charges brought against the defendants went back

[262] Eaton, William J. *"6 Go on Trial in Chernobyl Disaster: Former Chief of Nuclear Plant, 5 Aides Face Prison Terms."* Los Angeles Times (Los Angeles), July 8, 1987. Accessed March 24, 2016. http://articles.latimes.com/1987-07-08/news/mn-2679_1_chernobyl-plant.
[263] Ibid. Page 303.

to the earliest days of the plant, when the test was supposed to have been conducted during commissioning, but also spanned routine disregard of safety regulations and failure to provide proper on-site training. Bryukhanov claimed to be unaware that the test had not been completed originally, or that it had been planned for that night - whether or not that is true, we will never know - but he accepted that training and safety was substandard. Laushkin, the safety inspector, was charged with criminal negligence for repeatedly overlooking breaches of safety regulations and signing off on the test program without even looking at it. On that fateful night, a representative of the department of nuclear safety should have been present, and Dyatlov should have sought approval from the USSR's highest-ranking scientific minds before going ahead.

The full transcript and presented evidence remain classified to this day, so we'll probably never know much of what transpired. However, Nikolaii V. Karpan, Deputy-Chief of the Nuclear Physics Laboratory at Chernobyl's Nuclear Safety Department, who attended the trial on his days off, later published a book containing an extensive transcript based on private notes he took. Other attendees also took notes, but the KGB confiscated those. I assume Karpan was allowed to keep his because of his position in the nuclear industry. From his document, it's clear the Chairman of the panel of judges had no interest in hearing about reactor defects. Scherbina and Legasov's original government commission had discovered these defects and concluded that the reactor was at fault, but the only sections of their report given any credence by the judges were those criticising the operators. So-called 'independent experts' were, in fact, handpicked men from the various Institutes responsible for the reactor's creation in the first place - the very same men who had a vested interest in seeing their work exonerated. Their claims that the operators were entirely responsible is unsurprising at best, farcical at worst. Witnesses and defendants drew attention to the RBMK's flaws multiple times, but these comments were repeatedly interrupted or dismissed. So, too, were remarks that operating documents and regulations made little mention of the unreliable nature of control-board instruments at low power, that operators had no way of knowing the reactor became unstable and prone to explosion at these low levels, nor that disabling critical safety systems was, in fact, permitted if approved by the Chief Engineer or his Deputies. Dyatlov fought the official story throughout the trial, but even he was quoted as saying, *"With so many human deaths, I can't say I'm completely innocent."* When the Court asked why the regulations had not warned of any possible dangers associated with low power operation, the expert response was that an explanation, *"was not needed, otherwise the operating regulations would become inflated."*[264]

It was well known (though never acknowledged in public) that problems inherent with the Communist system meant planning never fully worked, and that citizens of all professions and levels of hierarchy were forced to improvise to get things done. The men seen playing cards on-shift at Chernobyl, for example, were probably doing so precisely because they had no purpose at the station - they were there because the Communist system had assigned them an unnecessary job already being fulfilled by someone else, so they were left with nothing to do. None of this could be admitted with the eyes of the world watching, so the legal proceedings played out as if the USSR was a perfect society. Everyone in attendance, including the six

---

[264] Karpan, Nikolaii V. Trial at Chernobyl Disaster. Report. Kiev, 2001. Almost everything in this and the following paragraphs is from Karpan's document as it is the only source for the trial, other than for the opening and closing days.

accused, knew the trial was for show. One daring witness on the stand even openly stated as such: *"I have a feeling that all foreign media will report, and all Soviet [society] will learn, that the accident happened as a consequence of mistakes committed by personnel,"* he remarked. *"Of course, the personnel are guilty of the disaster, but not in the scope defined by the Court. We worked with nuclear hazardous reactors. We had no idea that the reactors were highly explosive."* Dyatlov concurred: *"What happened [at the trial] was what always happens in these cases,"* he said later. *"The investigation was carried out by the very people who were responsible for the faulty design of the reactor. If they had admitted that the reactor had been the cause of the accident, then the West would have demanded the closing down of all other reactors of the same type. That would have dealt a blow to the whole of the Soviet industry."*[265] Karpan's own subsequent analysis of the case summed up how one-sided the proceedings were in a single sentence (question and exclamation marks present in original text): *"The indictment referred to these* [reactor] *defects as 'some particularities and shortfalls peculiar to the reactor' which played 'their role' (!?) and contributed to the accident 'in some way' (?)."*[266]

When all was said and done, the judge declared that, *"there was an atmosphere of lack of control and lack of responsibility at the plant,"* and found all six men guilty of causing the accident.[267] It was determined that staff had not received sufficient training; safety inspections were lacking; the test program was badly written; proper approvals outwith the plant had not been sought; among the people who did approve the test, none of them read it properly or identified problems with it; Dyatlov, in particular,

[265] Dobbs, Michael. *"Chernobyl's 'Shameless Lies"* Washington Post (Washington). April 27, 1992. Accessed March 04, 2016.
https://www.washingtonpost.com/archive/politics/1992/04/27/chernobyls-shameless-lies/96230408-084a-48dd-9236-e3e61cbe41da/.
[266] Karpan, Nikolaii V. Trial at Chernobyl Disaster. Report. Kiev, 2001. Page 52.
[267] *"Chernobyl Officials Are Sentenced to Labor Camp."* The New York Times (New York), July 29, 1987. Accessed March 24, 2016. http://www.nytimes.com/1987/07/30/world/chernobyl-officials-are-sentenced-to-labor-camp.html.

breached regulations that directly contributed to the accident; Bryukhanov initially hid the extent of the accident from senior figures in Moscow. Perhaps most seriously, the plant's management failed to initiate the disaster plan, resulting in thousands of people receiving far greater than necessary doses of radiation.[268] Bryukhanov and Fomin, as the highest ranking men at the plant, were each sentenced to ten years in prison, Dyatlov received five, Kovalenko and Rogozhkin: three, and Laushkin: two. Bryukhanov and Dyatlov - who wrote a book telling his side of the story some years afterwards, in which he placed the blame almost squarely on the designers - were released from prison early due to poor health brought on by radiation exposure. Chief Engineer Nikolai Fomin was declared insane in 1990 and transferred to a psychiatric hospital. Astonishingly, after he recovered he was allowed to return to work at the Kalinin Nuclear Power Plant near Moscow.

---

[268] On this point I sympathise with the men, as it was definitely a case of "damned if you do, damned if you don't." They didn't evacuate out of fear of reprisal, which they were now being punished for, but if they *had* evacuated earlier without permission, they likely would've been punished for that too.

CHAPTER ELEVEN

Departure

I stow the hefty Nikon SLR in my rucksack and lay it on the roof; I'm tired of seeing this incredible place through a lens, I want to use my eyes. Over the years, I have visited and photographed many abandoned buildings, and all too often I've realised afterwards that I hadn't really, properly *looked*, because I was too busy hunting for shots through my camera. These days I make a conscious effort to aim for more of a balance between taking photos and absorbing my surroundings. There's no time to visit another location and I'd rather take in the sights, sounds and smells around me than frantically rush about like a lunatic in these final moments.

The school roof is low relative to other nearby buildings and trees in the vicinity, only about 4 storeys high, so I can't see too far in any direction. A greenhouse with all of its glass miraculously still intact, countless trees and a smattering of unremarkable, slowly crumbling residential blocks. Still, it's tranquil up here; all I hear is wind rustling the leaves of nearby trees and the faint but ever-present chiming of the distant pile-driver/bell. Katie and I sit in silence, trying to extend the moment as long as we can, but soon - too soon - it's time to go.

We return down bare concrete stairs to the top floor where we find Danny and Dawid, both of whom agree little remains in this school devoid of children. Happy that I made the right decision to spend my final moments the way I did, we begin our reluctant trek back to the bus, along the golden path littered with fallen trees. I feel weighed down, as if I've been irreversibly altered by my brief trip to the Zone in a way I cannot describe, and already I know it'll always be with me.

Try as I might, I can't imagine people living here. I've seen photographs of streets full of smiling families and brand new cars, of couples dancing in the Palace of Culture and buying TVs from the hardware store. This city is now so unrecognisable - for the most part - from the locations in those pictures, I feel like they must have been taken somewhere else. What were once wide open spaces between buildings are now a maze of overgrowth, sometimes so thick you can walk between two structures and not see them unless you look straight up above the tree-line. Even though the evidence of prior residents is all around me, I simply can't visualise it here and now.

Our journey through the Zone holds one final, fleeting stop before we step back aboard the Slavutych train: the famous white 'Pripyat 1970' sign which welcomes visitors to the city. In a city full of signs and murals, it's perhaps the most recognisable of them all. We gather before it for a group photo, the same way I've seen wedding parties do in old black and white images from before the evacuation, and that's that.

The following morning, we have a little free time before the bus leaves Slavutych. After savouring a final cup of Ukrainian tea, Katie and I scale a meagre, wobbly ladder - precariously

217

placed over a five-storey drop in the stairwell - to the roof of our building. The town is so green. Everywhere you look are grass and tall pine trees in thick groups between buildings, almost as if the town's designers just plonked an assortment of buildings down in a forest, and only chopped the trees where they physically had to accommodate each structure.

We stow our bags in the bus, and with another half hour to kill, I head to the Chernobyl memorial in one corner of Slavutych town square. The faces of all 31 who died during those first months, engraved in two rows of pitch-black stone either side of a colourful assortment of flowers, stare back at me. I recognise a few key players - Akimov, Toptunov, Pravik - and I'll become more familiar with many of the rest in time, though I don't know it yet.

In stark contrast to our interminable journey into Slavutych some 60 hours earlier, the atmosphere aboard the bus is subdued. There's little talking, most either sleep off their exhaustion from the previous few days or gaze out the window, lost in thought. Not far outside town, we drive past a weathered-looking man in khakis and boots, perched atop a mound of vegetables on an ancient wooden horse-drawn cart. The contrast is glaring. Here we have a vehicle that has been in use for millennia, just a few dozen kilometers from a nuclear reactor, one of the most complicated and precise pieces of technology ever devised. A machine of which even the basic concept was considered impossible to achieve by the world's brightest minds only a century ago.

As northern Ukraine's flat, featureless landscape passes my window in a blur, I can't stop thinking about the night of the accident. What if those turbines had been tested properly during commissioning? What if the national grid management hadn't delayed the experiment, and more experienced operators had carried it out? What if Dyatlov hadn't been so stubborn, so determined against all logic and reason to press on after the power drop? What if Akimov and Toptunov had held firm and resolute, refusing to continue? What if the others present in the control room that night had backed them up? Would an accident have happened at Chernobyl regardless? At another RBMK reactor in another country - Russia or Lithuania? The reactor's flaws were known only to a few, but they were a powerful few, with the influence to fix those flaws if they had wanted to. Obviously they had not, or it wouldn't have taken a global catastrophe for them to take action. Don't rock the boat. What if the firemen, plant staff or Liquidators hadn't been so selfless in their battle to contain the accident? Or the Bio-robots who, against all concerns of self-preservation, stormed the poisonous roof. What if the wind that day had been blowing south towards Kiev - a city of close to 3 million people - instead of north and east, across mostly uninhabited rural farmland? What if the Soviet Union's response to the disaster had been unhurried, reluctant and lacklustre, with more concern paid to the financial cost of the clean-up than containing the problem, as TEPCO's has been with Fukushima?

Two hours fly by as I ponder the possibilities, and before I know it we've arrived at our one and only stop before reaching Kiev - a firing range. Being from Britain, I've had no exposure to firearms, but I *have* always been curious to know what it's like to use one. You see heroes in action movies hit moving targets with little-to-no effort; can it really be that easy? After crossing a field littered with empty shotgun cartridges, I soon discover the answer is a resounding nope!

My weapons are two Soviet classics - a Dragunov sniper rifle and the iconic AK-47. When it comes to my turn, I lower myself onto a rickety wooden stool, rest the barrel of the

Dragunov on a pockmarked bench in front of me, and brace the butt against my shoulder. The rubber eyepiece isn't sitting flush with the scope - it's angled down slightly - forcing me to look up into it at an angle, despite my quick attempt to realign it. Still, I've seen enough movies to know the basics of how to use one of these: take slow, deep breaths, relax, exhale and squeeze - not pull - the trigger.

Bang! The roar of the tiny explosion within the rifle's chamber is deafening, even though my thick ear-defenders. *"Miss,"* announces Marek through our translator, as he peers down-range towards the embarrassingly close target, not even 50 feet away. I couldn't care less, I'm more interested in how it feels to use an instrument designed to kill people than hitting anything with it. I empty the Dragunov's magazine, hitting nothing but dirt. Although I can't see him standing beside me, I can tell the instructor is looking at me with a disinterested mixture of pity and resignation.

He passes me the Kalashnikov, the most famous and widespread gun ever made. In service since 1949 and with over 75 million in existence today, according to The World Bank, the AK-47 is used in almost 100 countries, has a deep, distinctive sound, and has become synonymous with war. Again, I hit everything except my target, but after everyone's taken a turn and we're asked if any would like to pay for a second use, I hand over more notes. I have one purpose in mind. Until now, we've been firing semi-automatic rounds. I want to empty an entire magazine on automatic, the way you always used to see in 80s movies. As anticipated, the weapon kicks wildly in my hand and I struggle to keep it level, metal flying across the field. This time I know I won't strike the target, and I don't. I'm not at all surprised that untrained fighters always suffer greater losses than skilled soldiers; despite invariably firing more bullets, only blind luck will land a hit when firing more than a short burst.

It's early afternoon when Kiev's hazy, angular skyline creeps up over the horizon ahead of us. We're staying at the city's largest and most cringe-worthy-named hotel, the Tourist Hotel, a few hundred meters west of the mighty Dnieper River. The check-in process is quick and we eagerly head to our rooms to enjoy the view. It's magnificent. I snatch my camera from its bag and have to restrain myself from running out to the corridor, where I discover that almost every single one of us has collectively had the same idea: head to the roof. This is a little delusional, and we're soon forced to accept that it isn't happening - the requisite doors are locked, of course - so most people drift back off to their rooms. I'm a little more determined. I stalk the top floor corridors, soon finding a glass balcony door that, to my surprise, opens when I turn the handle. As I cross the threshold I'm silently thankful that Ukraine hasn't fallen victim to Britain's health and safety obsession.

The view is awe-inspiring, genuinely one of the most memorable moments of my life. The mid-afternoon Sun hangs low in the October sky, draping every bare-concrete structure and autumnal tree all the way to the horizon in a hard but warm light. To my left, a distant factory chimney puffs out white lines of rising smoke, standing out against the darkening sky. Off to my right, a busy main road passes the hotel before crossing over the Dnieper and Desenka Rivers, past forested islands. Ahead, the Motherland Monument is haloed in front of murky clouds as she stands vigilant over the city, her sword and shield raised. I feel a sudden urge to pay her a visit. Looking down over the worryingly thin, waist-high barrier, I spot several of my compatriots taking photos from five floors below me. I shout, they look up and laugh, giving me a great photo.

Throughout the afternoon, I make several trips up and down the building, sometimes with my friends, otherwise alone on the top floor balcony. At one point I bump into an Australian who works from a beautiful corner office on the top floor. We chat for a while about the city, Ukraine and Chernobyl. He says he loves working here; he travels a lot and usually spends a year or two living somewhere before moving on to the next place. I envy him. Before returning to my room, he urges me to go out into the city tonight while I'm here, and with that my mind is made up.

Dawid, Katie, Danny and I spend the next few hours sitting beside our open hotel window, watching the sunset and listening to all the sounds of the city. After dusk we take the elevator down to the lobby, turn right outside the entrance and walk down the bustling street towards the Dnieper River. After a hundred meters, the road nears a short bridge. Here, we leave the other pedestrians behind and turn onto a steep dirt path through a small wood, all the way to the river's edge. We take our photos then decide to sample Kiev's local cuisine, in the form of a nearby McDonald's. After a strict diet of cucumber, tomato and chicken, I'm desperate to eat something familiar. Cheap and greasy, but familiar. One awkward conversation and a Big Mac later, and we're back out in the night air. My friends want to spend the night recovering in their rooms, but, as exhausted as I am, I'm determined to see the city. She takes some convincing, but I manage to persuade Katie to accompany me.

First stop is the Patriarchal Cathedral of the Resurrection of Christ, walled off at the centre of a construction site not far from our hotel. Apparently it opened six months ago, but the grounds are unfinished; bits and pieces of machinery, masonry and equipment lie around the floodlit area. Katie and I find a secluded, shadowy corner and scale the wall, landing delicately, then take photos of the impressive white, gold and green church. We test the doors for a sneaky look inside, but they're all padlocked shut, so we leave and make our way back to the 700-meter Metro Bridge. It's a combined pedestrian, road and rail bridge, built of concrete in the 60s, as Soviet as it is exposed to the bitter October wind. Katie and I chatter (and shiver) as we cross, occasionally interrupted by the rattle of blue and gold metro trains, their lights flashing by us.

On the other side, we first enter the Metro station, but don't know if it'll take us up the Perchesk Hills, above the Dnieper's western bank. Instead, lacking a map, we opt for the Neanderthal method and climb - often on our hands and knees - up a small but steep and decidedly un-pedestrianised patch of woodland. Our only illumination: the dancing shadows cast by cars above and below. We're both reasonable climbers, used to taking obscure and often dangerous routes into abandoned buildings, and soon reach a gently inclined, perpendicular road without difficulty.

We follow a set of shallow steps leading off the road and up to a memorial park, then head north down a wide deserted street just beyond. Near an intersection at one end, we find ourselves outside the dainty, rotunda Church of St Nicholas. The orange-and-white neoclassical building is over 200 years-old. Katie and I take a few photographs, then return up the road into the multi-layered Park of Eternal Glory, beautifully lit by lamps of varying heights along each snaking footpath, and make straight for the impressive monuments on display. The first we reach is The Monument of Eternal Glory at the Grave of the Unknown Soldier, a 27-meter granite obelisk with an eternal flame burning at its base. The monument celebrates those countless unidentified soldiers who fell in battle during the Great Patriotic

223

War (what the Eastern-Bloc countries called World War II) of 1941-1945. Its flame warms us while we rest - we've already walked 5km. Nearby is the Candle of Memory, an intricate monument to the victims of the Holodomor, the genocidal man-made famine of 1932-1933 that killed up to 7.5 million Ukrainians. It's a modern and visually striking memorial: a 30-meter glass hexagon with hundreds of small crosses cut out of white panels on each side, all the way to the top. Four huge, backlit metal lattice crosses ring the base, while a glowing symbolic flame tops it.

It's after 9pm as Katie and I step out of the almost-deserted park and onto Lavrska Street, heading south in the vague direction of the Motherland Monument. We're soon walking parallel to the 20-foot wall surrounding the stunning Kiev Pechersk Lavra, a 280-year-old white and gold Orthodox Christian monastery (somewhat ironically sited opposite one of the country's oldest arms factories, the Kiev Arsenal). The intricately painted and engraved entranceway - depicting Angels and Holy men - is locked due to the late hour, leaving us no choice but to pause to appreciate the artistry, take a quick photo and move on. We pass by countless other examples of wonderful eastern European architecture.

I find myself both wishing we had more time and wondering what, in wake of the Chernobyl disaster, Kiev's residents thought of the military vehicles' daily radiation cleansing of the city. Testimonies from the time suggest a contrary mixture of fear and calm in the population. Fear, because of the disturbing rumours coming from Chernobyl, because Kiev's men were being woken from their beds at night and driven off to the zone, and because the KGB confiscated all dosimeters from laboratories across the city. Calm, because of the constant reassurances from all levels of government that the situation was under control and there was nothing to fear. It's known that all trains departing the city were fully booked - creating a black market for tickets - and the city's banks ran out of money on May 6th, after journalists were finally allowed to mention what had happened, suggesting panic on a large scale. Throughout the walk, I think often of the 1986 May Day celebrations, when these streets were flooded with tens of thousands of uninformed men, women and children of all ages. Some Communist Party officials - men who were aware of the true dangers - even paraded their own children around in a vain and selfish parody of normality. For those first months, futile attempts to avoid a mass panic took precedent over people's lives.

As we pass through a silent, unoccupied pedestrianised area that might be a marketplace during the day, we hear raised voices close behind us. Neither Katie nor I can see anyone else here, so we turn to find two uniformed policemen striding towards us. It seems they don't initially realise we're tourists, as one man speaks for a few sentences while gesturing at my bulky Manfrotto tripod. They're obviously not happy about it. Perhaps from a distance, in the dark, they thought it was a weapon? We try our best to explain, through subdued gesturing of our own, that we're tourists photographing the city and aren't trying to cause trouble. I wonder for a few seconds if Katie and I are about to be detained, but the pair decide we aren't worth the effort and wave us on.

I do a double take when a small collection of Soviet armoured vehicles appears around a corner, parked at the edge of a low hill. That's something you don't see every day: tanks sitting out in the street. At first I only see 6 tanks and APCs, including the T-54, T-55 and T-62 main battle tanks and a pair of BMP-1 and BMP-2 infantry fighting vehicles, but as I

225

progress along the line, the area opens to reveal more treasures. There's a distinctive 'Shilka' 4-barreled anti-aircraft tank, a PT-76 amphibious light tank, a 'Gvozdika' self-propelled howitzer, and the venerable Mil Mi-24 'Hind D' helicopter gunship - a childhood favourite. The area isn't well lit, with only a dim, amber streetlight on one side and the Moon on the other illuminating the scene, requiring long exposures of 30 seconds to capture a proper image. Around the next corner we stumble upon a glorious sight: a large open-air section of the Museum of the Great Patriotic War. Propeller planets; jet planes; tanks, big and small; armoured cars; missiles, and even an armoured train are on display.

The eclectic display is protected by that most effective of deterrents: a knee-high chain, strung between two low wooden posts. Katie and I step over it without hesitation and gleefully explore the assembled collection, which I guess encompasses a period from World War II to the 1970s. The piece that most draws my attention is a towering olive-drab and grey Lisunov Li-2 - Russia's own licence-built version of the classic twin-engine Douglas DC-3 transport plane. Behind it stand a selection of famous fighter aircraft. Three jets: a 1952 MiG-17, a 1959 MiG-21, a 1970 MiG-23, and a propeller-driven Yak-9 from WWII, the most mass-produced Soviet fighter of all time, with 16,769 built between 1942 and 1948. Nestled into a corner, in amongst a variety of Soviet tanks, self-propelled guns and missiles, is the armoured train, with a tank turret at either end.

At last we reach the Motherland statue we've been closing on all night - she's worth the wait. The silver figure holds both outstretched hands in the air atop a hill overlooking the city. Her stainless steel sword is slightly stubby after being shortened, when it was discovered that its full height made the statue taller than the cross on the nearby Kiev Pechersk Lavra. Nevertheless, its tip still reaches an imposing height of 102 meters above me. The mighty 13 by 8-meter shield in her left hand is emblazoned with the State Emblem of the Soviet Union, while her pose and general demeanour remind me of the Statue of Liberty. She was built before the accident, and I can't help wondering what those silver eyes have seen. The statue literally stands upon Kiev's Museum of the Great Patriotic War, and in front of her, a short way down the hill, there's an expansive parade ground. It's deserted, apart from us and two tanks facing each other in the near distance, their barrels crossed. There's not a cloud in the sky, meaning the entire scene - save for a few sparse spotlights - is illuminated by an ethereal light beaming down from the Moon and stars. Another perfect moment.

Katie and I walk around the parade ground for a while, photographing the tanks, sculptures and our vantage over the city, saying little. I have a burning desire to know what she's thinking, how she feels, but I don't ask. Satisfied that we've had a taste of this wonderful city, we retrace the 6km-or-so back to our hotel in almost complete silence.

I sleep like a dog, never have I felt so tired. The next morning passes like a blur. I've no idea who ordered it, but after a light breakfast an aging, black, stereotypically-ex-Soviet taxi collects our foursome from the hotel to drive us to the airport. Apart from our motorway drive to the hotel yesterday evening, this is the first time any of us have travelled through Kiev in daylight, so I perch on my seat with eyes glued to the window. As the city passes me by, emotions I've fought to suppress for days overwhelm me, and I weep silently into the glass. It's stupid, embarrassing, I don't even know why I'm doing it. I hide my face in shame. I was never impacted by the accident in any way, yet this trip has permanently changed who I am, leaving an indelible mark, and I know I'll never forget it. And I haven't. Not a day goes by when I don't think about that place and the people whose lives were destroyed by what happened.

CHAPTER TWELVE

Consequences

The disaster was the first major crisis to occur under the fledgling leadership of the USSR's most recent General Secretary, Mikhail Gorbachev. He chose not to address the public for three weeks after the accident, presumably to allow his experts time to gain a proper grasp of the situation. On May 14th, he announced to the world that all information relating to the incident would be made available, and that an unprecedented conference would be held with the International Atomic Energy Agency (IAEA) in August at Vienna. Decades of information control proved difficult to cast off in such a short time, however, and while the report *was* made available in the West, it was classified in the Soviet Union. This meant those most affected by the disaster knew less than everyone else. In addition, although the Soviet delegation's report was highly detailed and accurate in most regards, it was also misleading. It had been written in line with the official cause of the accident - that the operators were responsible - and, as such, it deliberately obfuscated vital details about the reactor.

Sceptical global experts attending the Vienna conference questioned Dr Valerii Legasov and his fellow scientists about the event for three hours, at the end of which they accepted his description with a standing ovation. It was a political triumph. However, it transpired that, *"members of the Soviet delegation were strictly instructed not to meet with foreigners [in private], not to answer any questions on their part, and to follow the published report in every respect. Only because of the resolute stand taken by Legasov was it possible to go away from this policy."*[269]

Legasov had his faults, but fundamentally he was a good, conscientious man who pushed against, and felt guilty about, both his own inaction before the disaster and the official, half-honest approach he was forced to adopt afterwards. It was too late. Both the pretence and his willingness to criticise a Soviet system that had gestated a belief in invulnerability had damaged his reputation. While reporting to his peers at the Soviet Academy of Sciences in October 1986, he stated, *"I did not lie in Vienna; but I did not tell the whole truth."*[270] Legasov decided to take a firm stance against the official explanation and penned several papers on the subject. In them, he criticised the underlying problems with the RBMK, the poor quality of training for nuclear operators, the complacency entrenched within the Soviet scientific community and nuclear industry in particular (one plant director was quoted as saying that a nuclear reactor is like a kettle, *"and much simpler than a conventional plant."*[271]), and proposed further research into safer reactor types. The papers fell under

[269] Shlyakhter, Alexander, and Richard Wilson. *"Chernobyl: The Inevitable Results of Secrecy."* Public Understanding of Science 1 (July 1992): 254. http://pus.sagepub.com/content/1/3/251.abstract Quoting Kalugin, Dr. Priroda, November 1990. (Soviet popular science magazine.)
[270] *Andrei Sakharov: Facets of a Life.* Gif-sur-Yvette, France: Atlantica Séguier Frontières, 1991. Page 657.
[271] *"Legasov Suicide Leaves Unanswered Questions."* Nuclear Engineering International. Accessed March 25, 2016. http://www.neimagazine.com/features/featurelegasov-suicide-leaves-unanswered-questions/. I have seen this quote in an old book of mine, but I can't find it now.

the purview of the KGB and were all either censored or not published at all.[272] [273]

With his reputation in tatters, his health ravaged by radiation absorbed at Chernobyl, his disillusionment with his country's unwillingness to focus more on safety, and feeling the weight of so many dead on his shoulders, Valerii Legasov hanged himself on the disaster's second anniversary - a day after his proposal for a reformed Soviet scientific community was rejected. During the hours preceding his death, he dictated his memoirs in the form of a lengthy voice recording, in which he concluded that the accident was the, *"apotheosis of all that was wrong in the management of the national economy and had been so for many decades."[274]* Some speculated that he was silenced for speaking out against the Soviet nuclear industry's safety record, forcing the government to investigate his death, but no foul play was ever officially acknowledged. On September 20th, 1996, then-Russian President Boris Yeltsin posthumously conferred on Legasov the honorary title of 'Hero of the Russian Federation' for the, *"courage and heroism,"* shown in his investigation of the disaster.

September 29th, 1986 marked the first time a reactor at Chernobyl was restarted since the disaster. Unit 1 was brought to, *"the minimal controllable level,"* according to government newspaper *Izvestia*.[275] All was not well, but after further repairs and a successful second restart on October 20th, the reactor was

---

[272] *"Chernobyl: Valery Legasov's Battle."* In Chernobyl: Valery Legasov's Battle. TV-Novosti. 2008. A lot of information for this and the subsequent paragraph come from this Legasov documentary.

[273] Mould, Richard F. *Chernobyl Record: The Definitive History of the Chernobyl Catastrophe.* Bristol, UK: Institute of Physics Publishing, 2000. Chapter 19 - The Legasov Testament. This chapter is an English translation of Legasov's memoirs. Many parts of it are used as a source throughout this paragraph and the rest of the book.

[274] *"Legasov Suicide Leaves Unanswered Questions."* Nuclear Engineering International. Accessed March 25, 2016. http://www.neimagazine.com/features/featurelegasov-suicide-leaves-unanswered-questions/.

[275] *"Moscow Reports Restart Of a Chernobyl Reactor."* The New York Times (New York), September 30, 1986. http://www.nytimes.com/1986/09/30/world/around-the-world-moscow-reports-restart-of-a-chernobyl-reactor.html.

subsequently taken to full power. There had been a critical shortfall in electricity supply in Ukraine following the accident, and the government felt it had to make Chernobyl operational as soon as possible. Unit 2 soon followed suit, but Unit 3 required serious repairs and was not restarted until December 4th, 1987.

Following the Vienna conference, the myth that operating staff had been more or less completely responsible for destroying Unit 4 was propagated for several years, both by the Soviet Union and experts from the International Atomic Energy Agency. A 1991 report by a Russian commission of experts to the USSR State Committee for the Supervision of Safety in Industry and Nuclear Power painted a different picture, and revealed that the information released to the IAEA in 1986 and 1987 was missing numerous vital facts. In an unusual move for the USSR, the report was scathing of the reactor's design and contained many complaints, including: *"as a result of the misguided selection of the reactor's physical and design characteristics by the designers, the RBMK-1000 reactor was a dynamically unstable system with regard to power and steam quality perturbations;"* *"The obvious discrepancy between the actual core characteristics and the projected design values was not adequately analysed, and consequently it was not known how the RBMK would behave in accident situations;"* *"For a number of the most important parameters, violations of which on 26 April 1986 were considered by the reactor designers to have played a critical role in the initiation and development of the accident, no emergency or warning signals were provided in the design;"* *"There are grounds to believe that the reactor designers did not undertake to assess the effectiveness of the Emergency Protection System in the possible operating modes;"* *"The designers and authors of the standard operating procedures for the RBMK-1000 reactor did not inform the personnel about the very real danger of a number of reactor characteristics,"* and perhaps most damning: *"the Commission considers it necessary to stress that all the design deficiencies of the [control rods] were, in fact, known*

231

before the accident." It went on and on - dozens of critical safety violations were present in the design. The report concluded that, *"the Chernobyl accident, which was initiated by the erroneous actions of operating personnel, had disproportionately disastrous consequences because of deficiencies in the design of the reactor."*[276]

The 1991 report to the USSR State Committee for the Supervision of Safety in Industry and Nuclear Power also touched upon another critical issue - that there was no consistent top-level responsibility for the nuclear industry at all, which was one reason a dangerous reactor like the RBMK-1000 was approved for production. *"All those involved in the development and operation of nuclear power plants are responsible only for those parts of the job which they perform themselves. According to international standards and practices, this overall responsibility should be borne by the operating organisations. So far, the USSR does not have any such organisations. Their functions of making the most important general decisions concerning a plant as a whole were and are usually performed by the corresponding ministries, which are government authorities. As a result, the decision making is separated from the responsibility for the decisions. Moreover, following the repeated reorganisation of government authorities, those bodies which made crucial decisions earlier no longer even exist. As a result, there are dangerous facilities for which no-one is responsible."*[277]

After the report's release, opinions within the scientific community shifted. New information exonerated plant staff of much of the blame; proved that they had not violated operating procedures as much as was previously claimed; that key areas of the reactor's documentation were inadequate; and that defective reactor design had played a significant role in causing the

---

[276] International Safety Advisory Group. *The Chernobyl Accident: Updating of INSAG-1: INSAG-7.* Vienna: International Atomic Energy Agency, 1992. Annex 1, Report by a Commission to the USSR State Committee for the Supervision of Safety in Industry and Nuclear Power. "*Causes and Circumstances of the Accident at Unit 4 of the Chernobyl Nuclear Power Plant on 26 April 1986.*" Moscow, 1991. All included quotes come from this report.
[277] Ibid. Page 87.

disaster. In 1992, the IAEA's International Nuclear Safety Advisory Group revised their original report to include the new information and published it as '*INSAG-7*'. This new report made it clear the accident would never have occurred had there been a proper culture of safety, feedback and oversight in the USSR's nuclear industry. Even though the official story remained that the operators *were* partially responsible, INSAG-7 reiterates the fundamental point that, *"Nuclear plant designs must be, as far as possible, invulnerable to operator error, and to deliberate violation of safety procedure."*[278] In total, the International Atomic Energy Agency identified 45 safety issues in their review of the Chernobyl power station after the accident: 19 of high severity; 24 medium; 2 low.

Major changes were made to the RBMK design, including improving the speed at which control rods entered the core during a SCRAM event, lowering the time for a complete insertion from 18 seconds to 12; reducing the positive steam void coefficient of reactivity, and the effect of reactivity if there was a complete void in the core; installation of a Fast Acting Emergency Protection system, complete with an additional 24 control rods; removing the ability to bypass emergency protection systems while the reactor was at power, and, most importantly, a new control rod layout with a longer boron section and no empty/water section ahead of it. The graphite tip remained.[279]

Despite international calls for Chernobyl to be decommissioned at once, it endured a very gradual demise. On October 11th, 1991, just five years after the Unit 4 explosion, there was a third major accident at the plant, this time at Unit 2.

---

[278] International Safety Advisory Group. The Chernobyl Accident: Updating of INSAG-1: INSAG-7. Vienna: International Atomic Energy Agency, 1992. Page 17.
[279] Ibid. Pages 27-28.

Prior to the event, the Unit had been taken offline following *another* accident - this time a fire in its section of the turbine hall, which had broken out during minor turbogenerator repair work. After extinguishing the blaze, the generator had been isolated and its turbine coasted down to about 150 rpm when a faulty breaker switch closed, reconnecting it to the grid. The turbine rapidly sped up to 3000 rpm in under 30 seconds, then, according to a 1993 report by the U.S. Nuclear Regulatory Commission, *"the influx of current to TG-4 overheated the conductor elements and caused a rapid degradation of the mechanical end joints of the rotor and excitation windings. A centrifugal imbalance developed and damaged generator bearings 10 through 14 and the seal oil system, allowing hydrogen gas and seal oil to leak from the generator enclosure. Electrical arcing and frictional heat ignited the leaking hydrogen and seal oil creating hydrogen flames 8 meters high, and dense smoke which obstructed the visibility of plant personnel. When the burning oil reached the busbar of the generator it caused a three-phase 120,000-amp short circuit."*[280] Firefighters rushed to the scene. All flammable material had been removed from Chernobyl's rooftops after the 1986 disaster, so there was little concern it would ignite, but the hall's fire ventilation systems couldn't cope with the heat and smoke. Fire teams realised the roof's support trusses, which had no fireproof coating and were unprotected by the sprinkler system, were in danger of failing in the heat. Despite their push to get extra water into the hall and up to the roof supports, the trusses failed and a 50 x 50-meter section of the roof collapsed when the pumps couldn't provide enough water for both the sprinklers and fire hoses.[281]

---

[280] *Information Notice No. 93-71: Fire At Chernobyl Unit 2*. Report. Washington D.C.: United States Nuclear Regulatory Commission, 1993.
[281] Ibid.

The reactor itself was undamaged, but extensive repairs would be needed to make the entire Unit operational again. Ukraine's new parliament decided to decommission the whole of Unit 2 instead. Unit 1's lifespan ended on November 30th 1996, in exchange for US$300 million paid to the government for the modernisation of Ukraine's power sector, including improvements to Chernobyl's remaining reactor. Despite this, the plant struggled through its final few weeks, during which it was forced to shut down first because of weather damage to electricity infrastructure and then from a steam leak. In a televised event on December 15th, 2000, Ukrainian President Leonid Kuchma ordered the permanent shut down of the plant live from Unit 3's control room, saying, *"To fulfil a state decision and Ukraine's international obligations, I hereby order the premature stoppage of the operation of reactor number 3 at the Chernobyl nuclear power plant."*[282] With that, Chernobyl's last reactor ceased producing power for the final time.

Most new Soviet nuclear plants in the planning or construction stages were either put on hold or cancelled altogether, while new, stricter safety regulations led several existing plants to be shut down for various reasons. By 1989, planned nuclear capacity had reduced by 28,000MWe (for comparison, Unit 4 at Chernobyl produced 1,000MWe and was the most powerful type of reactor at the time). The Government eventually scrapped all plans for future development of the RBMK design, beyond maintaining and improving those already in use. Other than those already under construction, no further RBMKs were ever built. Of the 17 RBMK reactors commissioned, 11 remain in operation today. Since the Chernobyl disaster, the Russian Government has exclusively

[282] *"BBC News | EUROPE | Chernobyl Shut Down For Good."* BBC News. December 15, 2000. Accessed March 25, 2016. http://news.bbc.co.uk/1/hi/world/europe/1071344.stm.

built VVER reactors - the same type that competed against the RBMK to begin with.

Official USSR Government figures state that 30 men and one female security guard died as a direct consequence of the accident. That list only covers the people at the site within the first few hours of the explosion who died from acute radiation syndrome or burns. It ignores all military personnel who died due to exposure from the clean-up operation, civilians living in the surrounding area, and many others who entered the Zone shortly after the accident (journalists, doctors etc). Those whose bodies were recovered are buried in welded zinc coffins, to prevent their radioactive remains from contaminating the soil.

Even though the health impact of the accident was (and still is) given almost unprecedented attention by world experts, *"The actual number of deaths caused by this accident is unlikely ever to be precisely known,"* according to a 2006 report of the UN Chernobyl Forum's Health Expert Group. The area of contamination is too large (approximately $150,000^2$ miles of land, covering 23% of Belarus, 7% of Ukraine, plus large areas of western Russia and some eastern European countries [more/less, depending on your definition of 'contaminated']), poor health brought on from exposure is so difficult to attribute directly to radiation, and many fatal health problems manifest themselves years or even decades later.[283] [284] Each new study on the probable number of deaths yields completely different results from the last.

Ordinarily, I would eliminate the lowest and highest numbers, and assume a rough estimate from somewhere in the middle of the predictions. The IAEA estimated around 4000 deaths, which appears to be on the very low end of the scale,

---

[283] *Health Effects of the Chernobyl Accident and Special Health Programs*. Report. Geneva: World Health Organisation, 2006.
[284] *Chernobyl Disaster*. Report. Minsk: Belarus Foreign Ministry, 2009.

and which I hesitate to accept, given how many previously healthy people are known to have died within a decade of their participation in the Chernobyl disaster. According to Nikolai Omelyanets, deputy head of the National Commission for Radiation Protection in Ukraine, a Government-established 'permanent independent collegial supreme scientific expert advisory and consultative body on matters of radiation protection and radiation safety of Ukraine', *"At least 500,000 people - perhaps more - have already died out of the 2 million people who were officially classed as victims of Chernobyl in Ukraine.* [Studies show] *that 34,499 people who took part in the clean-up of Chernobyl have died in the years since the catastrophe. The deaths of these people from cancers was nearly three times as high as in the rest of the population."*[285] He also claims his teams found that infant mortality - presumably within the contaminated zones - increased 20% to 30% because of chronic exposure to radiation after the accident. Evgenia Stepanova, of the Ukrainian Government's Scientific Centre for Radiation Medicine, said: *"We're overwhelmed by thyroid cancers, leukaemias and genetic mutations that are not recorded in the WHO data and which were practically unknown 20 years ago."*[286] A 2006 report titled The Other Report on Chernobyl (TORCH), though commissioned by those with anti-nuclear interests and hence questionable, estimates a more moderate 30,000 - 60,000 additional cancer deaths. The likes of Greenpeace seem to have picked the largest numbers they can find - sometimes into seven figures - and blindly assumed they're correct. Having judged the merits of the various reports and read their respective critiques, I personally think the number is perhaps somewhere around 10,000, but I want to stress that this is a completely unscientific

[285] Vidal, John. *"UN Accused of Ignoring 500,000 Chernobyl Deaths."* The Guardian. March 25, 2006. Accessed March 25, 2016.
http://www.theguardian.com/environment/2006/mar/25/energy.ukraine.
[286] Ibid.

guesstimate. I just have trouble accepting a number as low as 4,000 when there's so much empirical evidence that disputes it. There is not a single report that isn't disputed in one way or another, so we will never be certain.

Of course, fatalities are only part of the story, as a vast number of survivors continue to suffer from significant health problems brought on by exposure to radiation. Reliable health statistics from before 1986 have been difficult to uncover, making comparisons problematic, but cases of birth defects, congenital deformities and leukaemia in children all appear to have risen sharply within five years of the accident. *"In the 30 hospitals of our region* [Rivne; 500 kilometers west of Chernobyl]*, we find that up to 30% of people who were in highly irradiated areas have physical disorders, including heart and blood diseases, cancers and respiratory diseases. Nearly one in three of all the newborn babies have deformities, mostly internal,"* said Alexander Vewremchuk, of the Special Hospital for the Radiological Protection of the Population in Vilne in 2006.[287] Even now, some hospitals in Belarus have signs inviting victims of Chernobyl to skip to the front of the queue. The New York Academy of Sciences has acknowledged a significant surge in all types of cancer, an increase in infant and perinatal mortality, delayed mental development, neuropsychological illness, blindness and diseases of the respiratory, cardiovascular, gastrointestinal, urogenital and endocrine systems in the affected areas.[288]

As with seemingly all post-accident numbers surrounding Chernobyl, it isn't known exactly how many people suffered non-fatal health problems from the accident. It *is* known that a

---

[287] Ibid.
[288] Yablokov, Alexey V., Vassily B. Nesterenko, and Alexey V. Nesterenko. *"Chernobyl: Consequences of the Chernobyl Catastrophe for the Environment."* Annals of the New York Academy of Sciences 1181, no. 1 (2009). http://www.nyas.org/Publications/Annals/Detail.aspx?cid=f3f3bd16-51ba-4d7b-a086-753f44b3bfc1.

lot of these people have been forgotten by their societies, and many workers from the Zone have found it impossible to leave. There's a social stigma surrounding people affected by Chernobyl; many companies won't hire them and people won't associate with them out of an ignorant and prejudicial fear of radiation. Some receive compensation from the government, but it's a tiny and dwindling figure. A number of those who returned to the Zone in the years following the accident specifically cite the difficulty they had being accepted elsewhere as a primary factor, even though it is still unsafe to live there. All over Eastern Europe, farms still cannot sell their produce because the animals feed off ground considered too radioactive and unsafe for consumption.

Chernobyl was a direct reminder to the nations of the world that nuclear weapons were too terrible to use. On May 15th, 1986, Dr. Robert Gale hosted the first press conference on the accident at the USSR's Ministry of Foreign Affairs. After making a statement on the patients' conditions, he spent a significant amount of time answering questions. One question in particular regarded what lessons would be learned from Chernobyl. *'I think we have to view what's happened these past few weeks in a broader context,"* he responded. *"We've been dealing with a relatively small accident and, even with international co-operation, our ability to respond and care for the wounded has been limited. If we have a difficult time in helping three hundred victims, it's obvious that any response to the intentional use of nuclear weapons will be inadequate. People who believe meaningful medical assistance is possible for the victims of nuclear war are mistaken."*[289] He voiced - in public - a fact that the Kremlin had no choice but to face, almost 40 years after the Cold War began. There was an accident plan for an event like Chernobyl, but one isolated

[289] Gale, Robert Peter, and Thomas Hauser. *Chernobyl: The Final Warning.* London: Hamish Hamilton, 1988. Page 86.

incident had stretched resources to their limit, showing the politburo for the first time what the consequences of a nuclear war would look like. One nuclear power plant had suffered what was a relatively small explosion - compared to a nuclear weapon - in a single reactor, and the Government was forced to mobilise the largest peacetime military force in history. Radiation had made any ordinary course of action impossible, pushing them to acknowledge that the use of even one nuclear bomb - let alone the 65,000 in existence in 1986 - was unconscionable.

Five months later, on October 11th, 1986, Mikhail Gorbachev met with US President Ronald Reagan to discuss the possibility of nuclear abolition. Both men agreed that something must be done and, on December 8th 1987, the Soviet Union and United States of America signed the Intermediate-Range Nuclear Forces Treaty to eliminate all land-based missiles held by the two states with ranges between 500 and 5,500km. Less than a year after the accident, the historic 'Forum For A Free World' conference was held in Moscow, attended by prominent figures from a wide variety of fields. This meeting of minds, combined with the lasting effects of the accident, helped persuade many hard-line Soviet politicians to finally accept that a nuclear war was unthinkable - unwinnable - and would destroy the planet. Global nuclear weapons tests declined in number. 1996 saw the creation of the Comprehensive Nuclear Test Ban Treaty, and physical testing came to a halt - to be replaced by computer simulation. That is until only two years later, when India and Pakistan both tested their own weapons in what were thankfully isolated events. Since then, the only nation to defy common sense has been North Korea.

Two years after the accident, the USSR acknowledged that the Chernobyl disaster had so far cost them 11 billion Roubles (at a time when a Rouble wasn't far off the value of a Dollar), while Gorbachev himself admitted a figure of 18 billion in 2006. That does not include a lot of secondary expenses, and even then it appears to be a significant under-estimation, based on a report released by the Belarus Foreign Ministry in 2009. It revealed that the Government there still spends roughly $1 million daily on the accident, and, *"damage caused by the Chernobyl disaster is estimated at some $235 billion. However, the overall amount of money that Belarus and the international community invested into the recovery amounts to just 8 per cent of the total damage."*[290] The cost was catastrophic for the Soviet economy, as were its cascading effects on the coal and hydro energy industries. Soon after this, the oil price crashed to around half of its previous value, damaging the economy still further. The accident gave Gorbachev the excuse he needed to remove many high-ranking military and political opponents to his more transparent vision for the Communist Party, helping to further usher in the era of *'glasnost'* - transparency. The USSR never recovered; Chernobyl is seen as one of the primary catalysts behind its collapse.

Most of the prominent figures in this book are no longer with us, including Anatoly Dyatlov, who died of heart failure in 1995. He maintained his innocence until the very end. In 1992, he reiterated that, *"I found myself confronted with a lie, a huge lie that was repeated over and over again by the leaders of our State and simple technicians alike. These shameless lies shattered me. I don't have the slightest doubt that the designers of the reactor figured out the real cause of the accident right away* [which is true; they did - A.L.] *but then did everything to push the guilt onto the operators."*[291] Viktor Bryukhanov is now 80 years-old and still has a clear memory of the accident at Chernobyl. *"There were no cowards or dodgers,"* he said in a 2011 interview. *"All were dedicated to the plant, loved it and defended it."* [292]

---

[290] *Chernobyl Disaster*. Report. Minsk: Belarus Foreign Ministry, 2009.
[291] Dobbs, Michael. *"Chernobyl's 'Shameless Lies"* Washington Post. April 27, 1992. Accessed March 04, 2016. https://www.washingtonpost.com/archive/politics/1992/04/27/chernobyls-shameless-lies/96230408-084a-48dd-9236-e3e61cbe41da/.
[292] Asaulyak, Maksym. *"Viktor Bryukhanov: I Could Have Been Sentenced to Death."* Kyiv Weekly. April 28, 2011. Accessed June 04, 2014.
http://kyivweekly.com.ua/pulse/theme/2011/04/28/164825.html.

CHAPTER THIRTEEN

The Road Ahead

The Sarcophagus was never intended as a permanent solution. Rather, the concern at the time was to erect a structure to confine the radioactive release as rapidly as possible. As a consequence, it only had an estimated life of around 20 years - a time frame long since expired. In 1997, a Shelter Implementation Plan funded by 46 countries and organisations for a replacement - dubbed the New Safe Confinement (NSC) - was set in motion, with an estimated cost of €2 billion. Construction finally began in 2011, around the time I visited the area. The NSC, an enormous, one-of-a-kind arch, 250 meters wide by 165m long, and weighing a colossal 30,000 tons, is being assembled from prefabricated sections at a special holding ground 400m west of Unit 4. The first half was completed at the end of March 2014. Both halves were joined together a year later, and all the external skin has been attached and internal work is progressing well as of April 2016. Though it was originally supposed to be in place over the Sarcophagus by 2005, funds were difficult to come by and the NSC is still being built as of early 2016. It's expected to be finished before the end of this year. Upon completion, the entire arch will be slid along purpose-built tracks over the existing Sarcophagus, a process that will take about two days. It will be the largest movable structure ever built. Unlike the original Object Shelter, this new confinement is designed to last 100 years, by which time most decommissioning work on Unit 4 should have concluded.

Each half of the arch is made from several sections. Enormous jacks, whose only prior job was raising the sunken Kursk submarine in 2001, were used to lift each stage higher and higher until it reached its full height of 110 meters. Inside are remotely-operated heavy-duty overhead cranes, to be used for moving people and equipment.

To avoid corrosion of the steel structure, the designers have implemented a clever air-conditioning system that circles 45,000m³ of warm air per-hour within the vicinity of the shelter's cladding. *"There are steel structures that have lasted 100 years, such as the Eiffel Tower, but they last because they're continually repainted,"* said Dr Eric Schmieman, a senior technical advisor from Pacific Northwest National Laboratory in the US, to Wired magazine in 2013. *"We're not able to do that once we slide this into place - the radiation levels are so high we can't send people in. So what are we going to do? We are going to condition the air that goes into that space. We're going to keep the relative humidity in there at less than 40 percent."*[293]

When everything is in place, engineers will begin dismantling the Sarcophagus - estimated to take five years. Assuming that's completed before 2023, when the Designed

---

[293] Hankinson, Andrew. *"Containing Chernobyl: The Mission to Diffuse the World's Worst Nuclear Disaster Site."* Wired UK. January 3, 2013. Accessed March 25, 2016. http://www.wired.co.uk/magazine/archive/2012/12/features/containing-chernobyl.

Steel Stabilisation Structure holding up the western wall is no longer guaranteed to take the weight, work can begin on removing fuel-containing material from within Unit 4. They'll have 100 years, which sounds like a lot, but nuclear decommissioning is a notoriously laborious process. Despite the fire at England's Windscale nuclear plant happening all the way back in 1957, clean-up work isn't expected to finish until 2041.

As for Fukushima, it is very much a man-made disaster - one with a post-accident story almost as interesting as Chernobyl's. Unfortunately, this is mainly because of how inept the clean-up operating has been. Every week for the first few post-tsunami years, new reports emerged of fresh leaks of radioactive water, decommission workers being exposed to high doses, inadequate equipment and safety precautions that would be considered laughable if they weren't endangering people's lives - not to mention the environment. They have even repeated the most frustrating error to occur in 1986: using radiometers that go off the scale and assuming the radiation levels are at the device's maximum rated measurement. Even more unbelievable, the clean-up operation has now become infamous for using unskilled homeless men and women, tempted off the streets by corrupt subcontractors who are often a barely-legal face for organised crime. These poor people work and live under horrifying conditions and have upwards of one-third of their money skimmed off wages by the same subcontractors who hired them.[294] Unlike the Chernobyl clean-up, where the Soviet Government threw men and money at the problem until it was buried, Fukushima's owner/operator Tokyo Electric Power Company (TEPCO) is a public company (though

---

[294] Slodkowski, Antoni, and Mari Saito. *"Special Report: Help Wanted in Fukushima: Low Pay, High Risks and Gangsters."* Reuters. October 25, 2013. http://www.reuters.com/article/us-fukushima-workers-specialreport-idUSBRE99O04320131025.

effectively nationalised in 2012 with a massive government bailout), with profits to make and investors to please. As such, it has spent as little money as it can reasonably get away with whilst giving the appearance of trying to resolve the problem.

In October 2013, Japanese Prime Minister Shinzo Abe ended a two-year period of stubbornly refusing international help, when he asked the world's nuclear experts for assistance in the clean-up. Mere weeks later, it was revealed that the Japanese Government had become so frustrated with TEPCO that it drafted a proposal to strip the company of its responsibility for the plant. At the beginning of November of the same year, already battling low morale, Fukushima plant operators began the most dangerous and delicate phase of decommissioning up to that point: removing highly radioactive spent fuel from Reactor 4's cooling pool. Japan's Nuclear Regulation Authority chief personally advised TEPCO's president Yoshimi Hitosugi to proceed with the utmost caution, but when asked for his thoughts on the matter, Hitosugi was blasé, insisting, *"we believe it's not dangerous."*

By March 2015, TEPCO had wasted more than a third of the $1.6 billion of taxpayer money allocated for cleaning up the plant in a catalogue of failures. A drastic plan to seal Fukushima Daiichi off from the surrounding earth, to stop contaminated water from leaking into the sea, was approved and the required machinery built. The joint TEPCO and government effort involves freezing the ground using 1568 pipes in a colossal wall 30 meters deep. Critics of the plan pointed out that cost and feasibility issues were not properly thought through, but the government pushed ahead with it anyway. An initial attempt to freeze the earth ended in embarrassing failure in 2014 when TEPCO couldn't get the temperature as low as required, even after adding ten tons of ice into the mix. Freezing was expected to resume in March 2016 - the original anticipated completion date - and will last 7 to 8 months if all goes according to plan.[295]

The biggest waste was a $270 million machine custom-built to extract radioactive caesium from water leaking out of Fukushima's three damaged reactors and into the ocean. The machine never worked properly and only filtered a total of 77,000 tons of water, instead of the 300,000 it was intended to process *every day* before it was abandoned. The leaking storage tanks mentioned above somehow cost $135 million, all of which are being replaced.[296]

Nuclear power had been experiencing something of a renaissance before the Fukushima disaster, with the world appearing to finally move on from Chernobyl. A fresh emergency brought old fears back to the surface, causing many countries to review their nuclear policies. For their part, Japan immediately shut down all 48 of its remaining nuclear reactors after the accident in 2011, and only recently began reactivating a select few. Germany, another major user of nuclear energy, followed suit and began decommissioning all of its plants, with Sweden, Spain and Italy each doing the same. Even France, famous for relying on nuclear power for about 75% of its electricity, appears to be shying away from atomic energy. The Obama Administration had encouraged construction of the first new US nuclear plants in decades, but these projects are already running over-budget and over-schedule. New technologies which could potentially change the tide, such as reactors using molten salt, are expensive and unproven on a commercial scale, with probable disadvantages often outweighing the theoretical

[295] *"Fukushima Daiichi Ice Wall Equipment in Place."* World Nuclear News. February 10, 2016. http://www.world-nuclear-news.org/RS-Fukushima-Daiichi-ice-wall-equipment-in-place-1002165.html.
[296] Yamaguchi, Mari. *"Japan Audit: Millions of Dollars Wasted in Fukushima Cleanup."* AP. March 24, 2015. http://bigstory.ap.org/article/75dd3b31041949b7bbd4de14a2d5b287/japan-audit-millions-dollars-wasted-fukushima-cleanup.

advantages, while many existing reactors are getting near the end of their intended lifespan and will soon be shut down forever. The nuclear industry - vital, yet feared and misunderstood - faces an uncertain future.

It isn't all bad, however. Even the staunchest opponents of nuclear power - the environmentalists - are now concluding in droves that it could be our only option for scalable, sustainable, clean energy, while India, South Korea, Russia and especially China are building over 60 new nuclear power stations between them. Exciting new technologies are being developed in India, where the world's first prototype commercial thorium reactor (which uses fission of uranium^{233}, produced from the natural element thorium) should be built by 2017. It can operate for four months without any human control and has been designed to last 100 years - triple the usual lifespan. Thoughts turned to tsunami-proof power stations after Fukushima, and now a team of nuclear engineers from MIT are working on a sea-worthy, floating reactor, which uses flooded compartments as an infinite supply of coolant. Competing renewable energy technologies like wind and solar are improving all the time, and may be a viable alternative to coal, oil and nuclear fuels in a few decades, but for the time being nuclear power seems like our only realistic chance of creating clean energy on a global scale. Let's hope that those with the power and money to build and run them have learned to put safety first.

22357617R00147

Printed in Great Britain
by Amazon